TS 16949:
Insights from a Third-Party Auditor with a Process Approach Audit Checklist

Also available from ASQ Quality Press:

The Process Approach Audit Checklist for Manufacturing
Karen Welch

*Process Driven Comprehensive Auditing: A New Way to Conduct
ISO 9001:2000 Internal Audits*
Paul C. Palmes

The Process-Focused Organization: A Transition Strategy for Success
Robert A. Gardner

The Process Auditing Techniques Guide
J. P. Russell

The Internal Auditing Pocket Guide
J. P. Russell

How to Audit the Process-Based QMS
Dennis R. Arter, Charles A. Cianfrani, and John E. (Jack) West

ISO Lesson Guide 2000: Pocket Guide to Q9001:2000, 2nd Edition
Dennis R. Arter and J. P. Russell.

Quality Audit Handbook, 2nd Edition
ASQ Quality Audit Division

Quality Audits for Improved Performance, 3rd Edition
Dennis R. Arter

The ISO 9001:2000 Auditor's Companion
Kent A. Keeney

To request a complimentary catalog of ASQ Quality Press publications,
call 800-248-1946, or visit our Web site at http://qualitypress.asq.org.

TS 16949:
Insights from a Third-Party Auditor with a Process Approach Audit Checklist

Karen Welch

ASQ Quality Press
Milwaukee, Wisconsin

American Society for Quality, Quality Press, Milwaukee 53203
© 2005 by American Society for Quality
All rights reserved. Published 2005
Printed in the United States of America

12 11 10 09 08 07 06 05   5 4 3 2 1

Library of Congress Cataloging-in-Publication Data

Welch, Karen, 1961–
    TS 16949: insights from a third party auditor with a process approach audit
    checklist / Karen Welch.
        p. cm.
    Includes bibliographical references and index.
    ISBN 0-87389-654-8 (alk. paper)
    1. Quality control–Auditing.   2. Process control.   I. Title.

    TS156.W453   2005
    658.4′013–dc22

                                                                    2005004128

Publisher: William A. Tony
Acquisitions Editor: Annemieke Hytinen
Project Editor: Paul O'Mara
Production Administrator: Randall Benson

ASQ Mission: The American Society for Quality advances individual,
organizational, and community excellence worldwide through learning, quality
improvement, and knowledge exchange.

Attention Bookstores, Wholesalers, Schools, and Corporations: ASQ Quality Press
books, videotapes, audiotapes, and software are available at quantity discounts
with bulk purchases for business, educational, or instructional use. For
information, please contact ASQ Quality Press at 800-248-1946, or write to ASQ
Quality Press, P.O. Box 3005, Milwaukee, WI 53201-3005.

Quality Press
600 N. Plankinton Avenue
Milwaukee, Wisconsin 53203
Call toll free 800-248-1946
Fax 414-272-1734
www.asq.org
http://qualitypress.asq.org
http://standardsgroup.asq.org
E-mail: authors@asq.org

AMERICAN SOCIETY
FOR QUALITY™

To place orders or to request a free copy
of the ASQ Quality Press Publications
Catalog, including ASQ membership
information, call 800-248-1946. Visit our
Web site at www.asq.org or
http://qualitypress.asq.org.

∞ Printed on acid-free paper

Contents

List of Figures and Tables

Preface

Finally, a comprehensive process audit checklist has been developed for use with ISO/TS 16949:2002[1]! This checklist does what many others do not: it groups the questions by process rather than by standard clauses, automatically guiding the auditor to conduct a process approach audit. This manual was developed to assist anyone involved with conducting or planning quality system audits including quality auditors, quality managers, quality system coordinators, management representatives, and quality engineers. In addition, potential auditees in any function or position should find the questions useful in preparing for an audit. The manual includes:

- A brief overview of the process approach

- Discussion of problem areas often found by third-party auditors

- The process audit checklist, and

- Forms to be used in conjunction with the process audit checklist to increase audit effectiveness

As a third-party auditor, I have seen genuine limitations in internal quality audit processes due to inexperienced internal auditors. Many of them tell me they just aren't sure what questions to ask. After all, most of them only audit once or twice a year. How could they be as effective as someone who audits professionally? Utilizing this checklist takes the guesswork out of the internal audit process. You get many benefits, including:

1. Questions written and grouped by a third-party professional auditor.

2. The tools needed to conduct a successful audit from start to finish utilizing a true process approach. By using the checklist and its appendices, your internal auditors will be required to

audit by process and perform follow-up in associated areas to maximize benefits.

3. An audit that prepares all levels and functions in the organization for a successful third-party process audit. I continually find persons surprised when I ask them questions such as, how do you measure your processes? They're often surprised when I ask for evidence of continual improvement in their process, especially the support processes. The main reason for this surprise is lack of training and understanding. However, if the internal auditors ask these questions, your management and staff would not be caught off guard during external audits.

4. The benefits of a process-based audit without hiring a professional!

Acknowledgments

First and foremost, thanks to my husband, Peter M. Malmquist, for his support and encouragement over the years. His patience and understanding allow me the opportunity to spend time on the road doing what I love—auditing and constantly learning.

I will always be grateful to Barbara and Lex Welch for teaching me to believe in myself.

I'd like to express my appreciation to ABS Quality Evaluations for their continued support. In addition, to every company who has endured one of my audits, I thank you. I have learned from each and every one of you.

And special thanks to Dean Nichols and Verne Eva for giving me the opportunity to work with the best at Ford Motor Company.

Introduction

The original goal of this publication was to produce an ISO/TS 16949: 2002 checklist that could be used to conduct a process-based audit. This checklist, illustrated in Chapter 3, is meant for use as a tool by trained auditors. Though it was the author's intent to cover the basic requirements of the technical specification, your organization may find the need to add questions to ensure all requirements are addressed for your quality system. In addition, unique questions from your organization's procedures and work instructions should be added to optimize your audit. You should modify the questions to fit the needs of your organization as well as adjust for your terminology. Also, I encourage you to review the forms illustrated as appendices following Chapter 3. You may find these forms to be the most beneficial section of the entire publication.

In addition to the checklist, this manual provides an overview of the process approach. It was not the author's intent to provide in-depth instruction on the process approach. Because there are many other references readily available that already thoroughly cover the subject, Chapter 1 is meant to briefly summarize the method rather than repeat information available elsewhere. This chapter also includes strategies for conducting internal audits.

Chapter 2 describes common errors found during third-party ISO/TS 16949:2002 audits. By being aware of these common mistakes, your organization may be able to avoid them.

1

The Process Approach: An Overview

What are your customer-oriented processes? What processes exist to support them? What metrics do you use to measure these processes? These are the fundamental questions that should be addressed to develop a process-based quality management system. It should not be difficult, and your system should be based on the way your company naturally does business (at least in the beginning). When you are defining your processes and their metrics, take advantage of the opportunity to optimize your overall system.

First, let's think about your customer-oriented processes. Although each organization is different and must define its own unique processes, typically, these processes would include:

1. Customer-related—receipt of customer requirements, contract review, order entry, and so on

2. Design (if you perform design or any part of it)

3. Production, including maintenance, inspection, calibration, and so on

Once these customer-oriented processes have been determined, you must determine their sequence and interaction. Although a flowchart is not mandatory, it typically is the best, easiest method to use. Thus, I recommend the development of an overall process flow indicating how the system flows. For best results, this top-level process flowchart should be developed by a cross-functional team that includes upper-level management. For each process, determine its inputs and outputs. Use arrows to indicate direction, clearly showing what step comes next. Also, don't forget to determine measures for each process. And for each measure, establish a goal that is achievable within the next year. Many times it is also helpful to determine long-term stretch goals as well. Figure 1 illustrates an example of a partial flowchart that includes some typical customer-oriented processes. It is not all-inclusive.

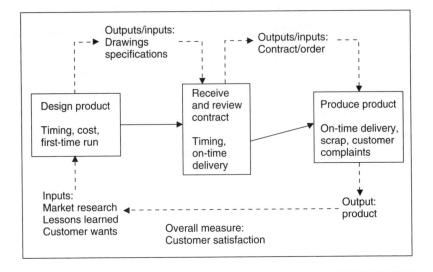

Figure 1   Partial example of customer-oriented processes.

Then, determine your support and management processes. Examples include:

1. Management review

2. Internal quality audits

3. Corrective and preventive action

4. Purchasing

5. Training

6. Document control

Add these processes to your flowchart in a manner to show how they support the customer-oriented processes. Also, it is important to include measures for each of these support processes as well.

Once this is accomplished, I strongly encourage you to take each customer-oriented process and support process and break it down further. Use cross-functional teams to develop an individual flowchart for each process. Although this is not required by ISO/TS 16949:2002, flowcharts are extremely useful tools to assist with process optimization. By clearly indicating the interactions required by different departments in the organization, the flowcharts can help identify and break down barriers that exist. We tend to operate with tunnel vision and focus on goals that optimize individual processes. Instead, we should work to optimize the process as a whole.

Why so much focus on metrics in the new ISO/TS 16949:2002? They are critical to your success. First, you must have measures to determine how well you are performing. Are you moving in the direction of your goals? If not, action plans should be developed to ensure that goals are reached. If you make changes to the system, will these changes be successful in helping you reach your goals? If not, should you be focusing your time and resources on the change? Second, employees will typically focus on what they are being measured by and/or what the organization is monitoring. Thus, it is critical for your company to establish the appropriate metrics. Otherwise, your people may be working on the wrong thing.

Now that you've defined your system and its processes, you must determine how to perform an effective process-based audit. It's very important to realize that each organization is unique, and effective audit techniques vary widely by organization and individual auditor. However, I'd like to share some basic strategies with you that I have found effective across different industries. The following strategies are discussed in subsequent sections of this chapter:

Section 1.1 Audit by Process

Section 1.2 Interview All Functions and Levels

Section 1.3 Encourage Top Management Involvement

Section 1.4 Welcome Nonconformities As Opportunities

Section 1.5 Focus on Known Weaknesses

Section 1.6 Thoroughly Train Your Auditors

Section 1.7 Allow Time for Auditors to Adequately Prepare

Section 1.8 Encourage Auditors to Put Auditees at Ease

Section 1.9 Use a Process-Based Audit Checklist

SECTION 1.1 AUDIT BY PROCESS

Without a doubt, auditing by process is a logical approach. However, it is not always easily done by new auditors or by auditors accustomed to auditing by element. It takes more knowledge of the overall system plus improved communication skills between audit team members.

A lead auditor should be assigned to coordinate the audit and develop an audit plan. The audit plan should identify the processes to be audited, dates, times, auditors, and so on. Be certain to allocate time for the audit team members to meet and discuss their status. Process audits rely on strong communication among auditors to ensure process linkages are adequately addressed.

To audit by process, ensure each requirement in ISO/TS 16949:2002 that applies to that process is covered. It is no longer as simple as auditing one element of the standard, as with QS-9000.[2] Each auditor must understand the entire technical specification and which sections apply to each process. Sections typically covered for each customer-oriented process as well as purchasing are:

4.2.3	Control of Documents
4.2.4/4.2.4.1	Control of Records/Records Retention
5.1.a/5.1.1	Management Commitment/Process Efficiency
5.3.d	Quality Policy
5.4.1	Quality Objectives
5.4.2	Quality Management System Planning
5.5.1	Responsibility and Authority
5.5.3	Internal Communication
6.1	Provision of Resources
6.2	Human Resources (all except 6.2.2.1 for specific to design)
6.3	Infrastructure
6.4	Work Environment
7.1	Planning of Product Realization (all)
8.2.3	Monitoring and Measurement of Processes
8.5.1	Continual Improvement
8.5.2	Corrective Action
8.5.3	Preventive Action

Tables 1–4 illustrate typical sections of ISO/TS 16949:2002 covered for some customer-oriented processes and purchasing, including those just listed. Keep in mind that your organization may need to add or subtract from these lists.

Table 5 illustrates typical sections that should be covered for the system overall and management. Again, your organization should review the list to determine whether or not it fits your needs. An auditor should be assigned to review the overall system, including the quality manual. Also, though documents and records should be sampled within each customer-oriented and support process, the overall process for document and record control should be observed. Similarly, the corrective and preventive action systems should be reviewed overall as a process as well as within each customer-oriented and support process. Also, the internal audit process should be reviewed overall for the company. The top management in the company should be audited for management responsibility and other areas noted in Table 5.

Please note that training requirements may be covered in each department and/or in the human resources department. The auditor must first determine area of responsibility as well as location of records. It is important to recognize that responsibilities may be shared by the individual departments and human resources. Either way, training should be recognized as a support process with its own process measure(s).

Prior to the audit of each process, the auditor should be familiar with the process flowchart and understand how the process to be audited links with other processes in the organization. Throughout the planning phase and the actual audit, the auditor should exchange information with the auditors working on other, linked processes. For example, Auditor A is reviewing production and Auditor B is reviewing purchasing/receiving. Auditor A finds a production work instruction that states all raw materials coming into the production area are prelabeled for traceability by the receiving warehouse. Auditor A should verify this with Auditor B to ensure this is in agreement with purchasing/receiving work instructions and actual practice.

Table 1   Customer-related process typical audit requirements.

4.2.3	Control of Documents
4.2.4/4.2.4.1	Control of Records/Record Retention
5.1.a/5.1.1	Management Commitment/Process Efficiency
5.3.d	Quality Policy
5.4.1	Quality Objectives
5.4.2	Quality Management System Planning
5.5.1	Responsibility and Authority
5.5.3	Internal Communication
6.1	Provision of Resources
6.2	Human Resources (all except 6.2.2.1)
6.3	Infrastructure
6.4	Work Environment
7.1	Planning of Product Realization (all)
7.2	Customer-Related Process (all)
7.4.1.3	Customer-Approved Sources (for suppliers)
7.5.4/7.5.4.1	Customer Property/Customer-Owned Production Tooling

Continued

Continued

Table 1   Customer-related process typical audit requirements.

8.2.1	Customer Satisfaction
8.2.3	Monitoring and Measurement of Processes
8.4.a	Analysis of Data—Customer Satisfaction
8.5.1	Continual Improvement
8.5.2	Corrective Action
8.5.3	Preventive Action

Table 2   Design process typical audit requirements.

4.2.3/4.2.3.1	Control of Documents/Engineering Specifications
4.2.4/4.2.4.1	Control of Records/Record Retention
5.1.a/5.1.1	Management Commitment/Process Efficiency
5.3.d	Quality Policy
5.4.1	Quality Objectives
5.4.2	Quality Management System Planning
5.5.1	Responsibility and Authority
5.5.2.1	Customer Representative
5.5.3	Internal Communication
6.1	Provision of Resources
6.2	Human Resources (all)
6.3	Infrastructure
6.4/6.4.1	Work Environment/Personnel Safety
7.1	Planning of Product Realization (all)
7.3	Design and Development (all)
7.6	Control of Monitoring and Measuring Devices (if applicable)
8.2.3	Monitoring and Measurement of Processes
8.5.1	Continual Improvement
8.5.2	Corrective Action
8.5.3	Preventive Action

Table 3   Purchasing process typical audit requirements.

4.1/4.1.1	General Requirements—Outsourced Processes/Supplemental
4.2.3	Control of Documents
4.2.4/4.2.4.1	Control of Records/Record Retention
5.1.a/5.1.1	Management Commitment/Process Efficiency
5.3.d	Quality Policy

Continued

Continued

Table 3 Purchasing process typical audit requirements.

5.4.1	Quality Objectives
5.4.2	Quality Management System Planning
5.5.1	Responsibility and Authority
5.5.3	Internal Communication
6.1	Provision of Resources
6.2	Human Resources (all except 6.2.2.1)
6.3	Infrastructure
6.4	Work Environment
7.1	Planning of Product Realization (all)
7.3.6.3	Product Approval Process
7.4	Purchasing (all)
7.5.3/7.5.3.1	Identification and Traceability/Supplemental
7.5.4/7.5.4.1	Customer Property (if purchasing assists with responsibility)
7.5.5/7.5.5.1	Preservation of Product/Storage and Inventory
7.6	Control of Monitoring and Measuring Devices (all)
8.2.3	Monitoring and Measurement of Processes
8.2.4	Monitoring and Measurement of Product
8.3/8.3.1	Control of Nonconforming Product/Supplemental
8.4.d	Analysis of Data—Suppliers
8.5.1	Continual Improvement
8.5.2	Corrective Action
8.5.3	Preventive Action

Table 4 Production process typical audit requirements.

4.2.3/4.2.3.1	Control of Documents/Engineering Specifications
4.2.4/4.2.4.1	Control of Records/Record Retention
5.1.a/5.1.1	Management Commitment/Process Efficiency
5.3.d	Quality Policy
5.4.1	Quality Objectives
5.4.2	Quality Management System Planning
5.5.1/5.5.1.1	Responsibility and Authority/Responsibility for Quality
5.5.3	Internal Communication
6.1	Provision of Resources
6.2	Human Resources (all except 6.2.2.1)
6.3	Infrastructure (all)
6.4	Work Environment (all)
7.1	Planning of Product Realization (all)
7.2.2.2	Organization Manufacturing Feasibility

Continued

Continued

Table 4 Production process typical audit requirements. *(continued)*

7.3.2.2/7.3.3.2	Manufacturing Process Design Input/Output
7.5	Production and Service Provision (all)
7.6	Control of Monitoring and Measuring Devices (all)
8.1	Measurement, Analysis, and Improvement (all)
8.2.1.1	Customer Satisfaction—Supplemental
8.2.2.3	Product Audit
8.2.3	Monitoring and Measurement of Processes
8.2.3.1	Monitoring and Measurement of Manufacturing Processes
8.2.4	Monitoring and Measurement of Product (all)
8.3	Control of Nonconforming Product (all)
8.4.b	Analysis of Data—Conformity to Product Requirements
8.5.1/8.5.1.2	Continual Improvement/Manufacturing Improvement
8.5.2	Corrective Action
8.5.3	Preventive Action

Table 5 Management process typical audit requirements.

4	Quality Management System (all)
5	Management Responsibility (all)
6	Resource Management (all)
7.1	Planning of Product Realization (all)
7.3.1	Monitoring (design)
7.4.3.2	Supplier Monitoring
8.1	Measurement, Analysis, and Improvement—General
8.2	Monitoring and Measurement (all)
8.4	Analysis of Data (all)
8.5	Improvement (all)

SECTION 1.2 INTERVIEW ALL FUNCTIONS AND LEVELS

It is important to interview all functions and levels during internal audits. This is what I commonly refer to as "spreading the joy." In other words, let everyone experience the audit process. Of course, unless you have a small organization, time does not permit everyone being interviewed at each audit. However, you should include as many persons as possible. Be certain to include persons in each department and at all levels, including the site manager, department managers, supervisors, operators, and so on. The organization should have auditors who feel comfortable interviewing operators as well as top management.

Beware of areas where only one person has all the answers. For example, if department managers developed the ISO/TS system for their areas, be wary if the managers try to dissuade you from talking to others who work for them. It is important that all personnel understand their role in the system. Granted, supervisors or operators may not have the answer to a question that the department is responsible for. Nor would they be expected to. However, they should be knowledgeable of their area of responsibility in the process. Thus, it is important for the auditor to cover all levels.

SECTION 1.3 ENCOURAGE TOP MANAGEMENT INVOLVEMENT

To optimize the audit process, top management should be involved and consider the audit a priority. Typically, all it takes is for the top manager to ask questions of the staff to obtain their participation. The auditor should not have to plead with personnel to set up the audit. Although the person developing the audit schedule should be flexible in working with the department managers, department managers should consider scheduling of the audit a priority.

Of course, top management involvement should apply to the entire quality system, not just the audit process. Without their buy-in and commitment, the system will not be optimized.

SECTION 1.4 WELCOME NONCONFORMITIES AS OPPORTUNITIES

It is critical for the organization to welcome nonconformities as opportunities. Nonconformities that are found during the audit should be viewed as quality system issues rather than personnel issues. Something in the system was not working properly, and the system must be corrected. If nonconformities are viewed as negative or if personnel are punished, it will be difficult to optimize the quality system. Personnel will not voluntarily offer information if they are fearful. However, if the organization views the audit as a means to improve the system, personnel will be much more likely to elaborate on questions that are asked of them.

The philosophy of the organization starts at the top; thus, top management must be involved, as noted in Section 1.3.

SECTION 1.5 FOCUS ON KNOWN WEAKNESSES

As we all know, the audit is based on sampling. With adequate preparation, the auditor should be able to greatly improve the benefit of the audit by focusing efforts on the most appropriate areas. The auditor should study the company's metrics, and during the audit, focus on areas with known weaknesses. For example, if trend charts indicate that one particular production area is creating the majority of the scrap, the auditor should focus on that area. Yes, other areas should be sampled as well. In fact, it is best to audit each and every production area. However, by spending more time on areas creating the most internal scrap and customer complaints, the auditor is best serving the organization.

SECTION 1.6 THOROUGHLY TRAIN YOUR AUDITORS

ISO/TS 16949:2002 requires the organization to have auditors qualified to audit the technical specification. Although TS leaves us with a great deal of flexibility on how to accomplish this, the organization that does the minimum to get by will probably not have a successful audit program. To optimize the audit process, every auditor should initially have formal training on the current version of ISO/TS 16949:2002, as well as auditing techniques. Also, when possible, it is best for a new auditor to witness other audits prior to becoming qualified to audit alone. It is not fair to the organization or the individual when adequate training does not occur. The organization loses because the audit process is not optimized. The auditors lose because they are put in a position that they are not ready for, and few people are comfortable in this situation. Ford Motor Company stresses auditor training in its customer-specific requirements for ISO/TS 16949:2002.

The process audit checklist illustrated in Chapter 3 is an excellent tool to assist a new auditor. However, it only defines questions for the auditor. It does not tell the auditor what to do with the response. The internal auditor must filter the questions to determine if the correct function is indicated for the function in the company. Thus, even with the checklist, training is critical for a successful audit. Ongoing, periodic refresher training for all auditors should be considered.

SECTION 1.7 ALLOW TIME FOR AUDITORS TO ADEQUATELY PREPARE

Many audits are not optimized because the auditors are not given adequate time to prepare. When auditors are assigned, it is important to build auditor prep time into the schedule. This allows the auditor's management to see how much time will be required. It is not enough to dedicate personnel's time just for the audit without considering preparation time. Because auditors may not audit their own work, many times they will be auditing in areas they are not familiar with. Preparation time is critical for the auditor to study documentation, trends in metrics, potential weak areas, and so on. It may be helpful for the auditor to prepare for the audit away from the normal workstation to avoid distraction.

SECTION 1.8 ENCOURAGE AUDITORS TO PUT AUDITEES AT EASE

To optimize the audit process, the auditor should try to put the auditee at ease. A nervous auditee sometimes makes mistakes and normally does not openly communicate. If the auditor can get the auditee to relax, in most cases, much more information will be obtained. Sometimes, if I suspect auditees are nervous, I will see if they have any photographs of pets at their workstations. If so, asking about the pet normally makes the auditee light up. (I choose pets because that is what I can relate to, but children work just as well.) The auditee tends to relax upon realizing that we have something in common. Of course, the auditor must always remain in control and stay on track. But taking 20 seconds to ask the auditee about the game last night will normally be worthwhile and beneficial to the audit.

SECTION 1.9 USE A PROCESS-BASED AUDIT CHECKLIST

Most auditors do not audit on a regular basis. In fact, many of them only audit a couple of days out of the year. It is often difficult for these auditors to determine the appropriate questions to ask. Although the process-based audit is much more beneficial to the organization, it is sometimes found to be more complex by new auditors or those who seldom audit. Thus, the use of a process-based audit checklist is highly recommended. The checklist in Chapter 3 is one example to consider. It should be considered a starting point, with each organization adding its own questions pertaining to unique procedures and work instructions.

2

Common ISO/TS 16949:2002 Errors to Avoid

As a third-party auditor, I continue to see common errors made over and over again during ISO/TS 16949:2002 Stage 1 Readiness Reviews and Stage 2 Initial Assessments. For the most part, it is quite easy to avoid these mistakes if the organization is armed with the correct information. This chapter does not cover the basics of ISO/TS 16949:2002 that are commonly available elsewhere. Instead, I pick up where most others leave off . . . with the difficult issues that a third-party auditor can help you with. From CEOs to quality mangers to external consultants, you will need this information to be fully prepared for your audit! Common mistakes I have found during third-party TS audits are given in the following list and discussed in subsequent sections:

Section 2.1   Unaware of All Automotive Customers

Section 2.2   Unaware of/Not Meeting All Customer-Specific Requirements

Section 2.3   Remote Locations Not Fully Disclosed

Section 2.4   Exclusions Improperly Listed

Section 2.5   Unaware of Customer Scorecards

Section 2.6   Linkages between Site and Remote Locations Not Clearly Defined

Section 2.7   Processes Not Fully Defined

Section 2.8   Measures Not Defined for Every Process

Section 2.9   Legibility of Documents Not Covered in Procedure

Section 2.10 Quality Objectives Not Defined at Relevant Functions and Levels

SECTION 2.1 UNAWARE OF ALL AUTOMOTIVE CUSTOMERS

It seems like a simple question: Who are your automotive customers? However, organizations sometimes fail to disclose all of their automotive customers during the Stage 1 Readiness Review. Some even go so far as to say they do not have any, not realizing this makes them ineligible for ISO/TS 16949:2002 registration.

The main problem seems to stem from a misunderstanding involving original equipment manufacturers (OEMs). Sometimes U.S. organizations fail to recognize that ISO/TS 16949:2002 is an international automotive standard; they believe "automotive customers" only applies to the U.S. Big 3 and mistakenly do not include their international automotive customers. More often, organizations disclose only the OEMs they supply, not realizing this requirement applies to other automotive customers as well. All major automotive customers should be revealed, including all OEMs and tier 1s. Though revealing "mom-and-pop" type automotive customers is not normally desired, it may be necessary if these customers are the organization's only link to automotive. Listing of lower tier customers is more commonly found among bulk material organizations. If you are bulk and you have hundreds of automotive customers, note the OEMs and tier 1s to your registrar, and let them know you have many more at lower tiers. If you know what tier your customers are in the supply chain, it may be helpful to note that for your registrar.

Sometimes the lack of disclosure of all customers is a result of lack of knowledge. Surprisingly, this sometimes occurs in larger organizations. The point of contact with the registrar for the audit may not be fully informed. Thus, both the external auditor and the company representative are surprised at the Stage 1 Readiness Review and sometimes the Stage 2 Initial Assessment audit. It is imperative for the company representative to ensure all accounts have been investigated to avoid this type of surprise.

In summary, disclose all major automotive customers to your registrar, not just U.S. OEMs.

SECTION 2.2 UNAWARE OF/NOT MEETING ALL CUSTOMER-SPECIFIC REQUIREMENTS

After the hurdle of identifying customers is conquered, the next important step is to identify all customer-specific requirements. At some point (typically during the Stage 1 Readiness Review), your registrar should ask for all these customer-specific requirements and the current revision date (see Table 6). Once again, this applies to all customer-specific requirements, not just the OEMs or U.S. companies. This list typically does not include contracts, terms and conditions of a purchase order, or letters/e-mails, and so on.

Ford, General Motors, and DaimlerChrysler each have separate customer-specific requirements especially for ISO/TS 16949:2002. It should be noted that these requirements are separate and different from their QS-9000 customer-specific requirements.

For other customers, customer-specific requirements include your customers' supplier quality manuals. Many customers have these, and I continue to be surprised by the number of organizations seeking certification that are not aware of these manuals. How can the organization accept a contract without first reviewing all customer requirements? That, in itself, is a significant nonconformity.

As of February 2005, a database to list customer-specific requirements, in my opinion, is much needed in the automotive system. It would be virtually impossible for anyone to know all existing customer-specific requirements; however, the third-party auditors will be aware of many of these, especially for the major automotive customers. It is up to the organization to investigate and determine which customers have these documented requirements. I have to presume that the organization prefers to know more about their customers than their auditor knows.

Table 6 Automotive customers/customer-specific requirements.

Customer	Name of customer-specific requirement	Revision date	Media	Scorecard?
U.S. OEM A	ISO/TS 16949:2002 customer-specific requirements	12/03	Online	Yes
U.S. tier 1 A	Customer-specific requirements	12/17/03	Hard copy	Yes
International OEM B	Supplier quality manual	Online	Online	Yes
International tier 1 B	Global supplier development and guidelines manual	7/03	Hard copy	Yes
U.S. tier 1 C	None	N/A	N/A	No
International OEM C	Supplier quality assuarance manual	4/10/03	Hard copy	Yes

Note 1: Include all major automotive customers, those with customer-specific requirements and those without. The list of major customers should include all OEMs and Tier 1's.
Note 2: This table is for illustration only. Check with your customers for current names and revision dates.

Surprisingly, once the customer-specific requirements are known by the organization, often they are not addressed. Why? Sometimes, the organization argues that the customer has never asked for the requirements stated in the manual. However, as an auditor, my approach must be that the customer believes the organization is meeting its requirements whether or not the organization specifically asks for each requirement. The exception here would be if the organization obtained a very clear waiver from the customer. If the organization should pursue a waiver, it is strongly recommended that the organization put the request in writing and ask the customer to sign it. Otherwise, the signed waiver may not be exactly what the organization needs if the customer does the documentation. This happens quite often (see Table 7).

Some organizations with a large number of customer-specific requirements seem to hope the auditor will not have time to cover everything. However, the organization should expect the auditor to sample from all these requirements. It has been my experience that it does not take very many questions to realize the organization is not familiar with the requirements. This is probably the one area where I find the most nonconformities overall.

What can the organization do to avoid these nonconformities? First, determine all customer-specific requirements documents. Second, as painful as it might be, go through each and every requirement and assign responsibility. The best-prepared organizations develop responsibility matrices to ensure all requirements are assigned and implemented (see Table 8). The organization will probably find many of these requirements repetitive with the standard; however, most customers will have additional requirements embedded in their manuals. And these additions will be the ones your auditor questions.

Table 7 Request for waiver.

The following scenario is an example of what to be cautious of when submitting a request for a waiver to a customer.

Customer: Company A

Organization (supplier to customer): Cookeville AutoMold

In reviewing Company A's customer-specific requirements, Cookeville AutoMold noted a requirement to submit certificates of analysis (C of A) one day in advance of the arrival of each shipment. AutoMold had supplied to Company A for five years, and they had never sent a C of A at any time. AutoMold called Company A and asked for a waiver. Company A readily agreed, but when they submitted the waiver to AutoMold, it was worded:

> "It is not necessary for AutoMold to submit a certificate of analysis one day in advance of the arrival of the shipment."

AutoMold was happy with this waiver until their third-party audit. At that point, it was pointed out that the phrasing received from Company A indicated they did not have to submit a waiver one day in advance, but it did not waive the certificate altogether.

Had AutoMold carefully worded the request and submitted it to Company A for approval signature, they may have avoided a nonconformity.

A good example of an unusual customer requirement recently found was the requirement for each supplier (in this case, the organization undergoing the third-party registration audit) to conduct an annual on-site audit of all its suppliers. Although this may have been somewhat common years ago, third-party audits have virtually eliminated this need (or so I thought). Often it is possible to get waivers for some of these requirements. Sometimes, customer contacts are not even aware of these requirements in their own company's supplier quality manual! It is the organization's responsibility to go through every requirement and ensure that it is meeting the requirements or has obtained waivers.

A very common customer-specific requirement is for the organization to submit a 24-hour contact list and/or organizational chart to the customer and to keep the customer up-to-date as changes occur. Though organizations often seem to submit the original request, the evidence to show it has been kept up-to-date is not common.

After the customer-specific requirements matrix is developed and completed, conduct an internal audit to verify that all is done. Keep the matrix as a living document for future changes and additions. By including all

Table 8 Customer-specific requirements—responsibility matrix.

Customer	Name of customer-specific requirement	Rev date	Page no.	Section no.	Summary of requirement	Responsibility– include names if possible	Fully implemented? If no, completion date
Company A	Company A supplier quality manual	Rev 3	1–2	A.1	All suppliers to Company A must conduct annual on-site audit of their suppliers	Purchasing	No—Nov 1
			2–3	A.2	Use AIAG APQP forms	Design	Yes
			4	A.2	24-hr contact list submitted to customer and kept up-to-date	Sales	Yes
Company B	Company B supplier quality assurance guide	Rev 1 11/03	1	I	SPC for all special characteristics	Manufacturing	No—Aug 15
			2–3	II	24-hr contact list submitted to customer and kept up-to-date	Sales	Yes
			4–6	III	Design verification record	Design	Yes
Company C	Company C guidelines for suppliers	Rev 2 9/00		I	Ensure all customer information is kept confidential	HR	Yes
				II	Return signed copy of terms and conditions	Sales	No—Jul 1

requirements on the original matrix (as opposed to only those that initially need clarification), this matrix becomes an excellent ongoing tool for your internal auditors.

An accredited registrar is required to list on a separate appendix to the certificate any customers whose specific requirements were included in the audit.[4] Thus, it is extremely important for the organization to correctly identify them. Don't forget to include heavy truck, motorcycle, and bus customers also. If you have any doubts about which customers to list, contact your registrar and discuss it up front.

SECTION 2.3 REMOTE LOCATIONS NOT FULLY DISCLOSED

Oh, those remote locations! The third-party auditor's nightmare—and often the organization's as well. Remote locations support the certified manufacturing site, and because they do not manufacture, they may not receive a stand-alone ISO/TS 16949:2002 certificate.

Folks, this is not QS-9000. What was okay for QS-9000 is not necessarily okay for ISO/TS 16949:2002. The rules are tougher. Keep in mind that the registrar's customer is your customer, not you, even though you are the organization paying the bill. Many, many companies are making the mistake of only listing the remote locations they had for QS-9000. While this is acceptable in some cases, it more often is not. This especially applies to large, multisite corporations. To be certified, the organization must be audited to all requirements of the technical specification wherever they are performed. The only two exceptions are allowable design exclusions (see Section 2.4) and some outsourced processes (see Section 2.34). Remote locations may include locations that perform design, purchasing, sales, customer service, warehouse, and so on.

For design, either the customer is responsible or the organization is responsible. Period. If the organization handles design at a technical center and/or corporate office, those sites must be audited. The organization may not exclude design just because the manufacturing site is not responsible. If design is outsourced, the registrar may determine a visit to that supplier is necessary.

Another common required remote location is one with some responsibility for purchasing. These sites often are responsible for selection of suppliers or writing purchase orders. And don't forget about those offices responsible for the purchasing of support services impacting the customer, such as carriers, warehouses, temporary agencies, and calibration services.

The largest problem area for remote locations involves customer requirements and the customer-related process. Often, companies have multiple sales and/or customer services offices where orders are taken. Often, especially with sales offices, specific customer and product requirements are taken and utilized in the design and quality planning processes. The lead auditor will most likely require these locations be audited as remote locations. These sales offices were not commonly audited with QS-9000, and organizations are commonly failing to disclose them to the registrar during contract negotiation and during the ISO/TS 16949:2002 Stage 1 Readiness Review. (An exception to this would be salespersons who work out of their homes; the auditor would not likely go to their homes to audit them.)

It is far better to disclose these remote locations or potential remote locations to your registrar in the initial stages of contract negotiation. The organization only has 90 days from the Stage 1 Readiness Review approval until the beginning of the last site audit. If an additional remote location has to be added after the audit has begun, the 90-day period is sometimes difficult to meet. Remember, registrars have limited ISO/TS 16949:2002 audit resources and scheduling must occur in time to allow these additional remote locations to become approved. If the 90-day deadline is not met, the entire audit process must start over.

Another common mistake the organization makes is to assume a sister plant seeking separate certification that supports the plant does not fall under the 90-day rule. It does. For example, I recently had a plant scheduled for May. Several weeks later, I received an audit confirmation for a sister plant seeking a separate certificate for October. The first plant listed the sister plant as a supporting location. Obviously, this would not work, and the audits had to be rescheduled much closer together. This makes it critical for large companies with plants that support one another to coordinate the audit schedule to ensure all can be completed within 90 days. The 90-day rule is not just for certificates issued under a corporate scheme. Remote support functions must be audited before the manufacturing site during initial assessments.[5]

In summary, disclose all potential remote locations to your registrar. If you question the need to include any of these sites, open a discussion with your registrar and/or lead auditor.

SECTION 2.4 EXCLUSIONS IMPROPERLY LISTED

Often companies identify improper exclusions. Only requirements in Section 7.3, Design and Development, for "product" design may be excluded.[6] The organization may not exclude process design. You may only exclude requirements of product design that you do not perform, and you must include justification in your quality manual. And as noted in Section 2.3, for automotive, either you or your customer must be design responsible.

Organizations tend to automatically assume that if they were certified to QS-9000:1998 based on ISO 9002:1994 that they may exclude product design. That is not always the case. They may have chosen to pursue only ISO 9002:1994 as opposed to ISO 9001:1994, even though they performed product design. With ISO/TS 16949:2002, you may not exclude product design if you do it. And in addition, if you perform any part of product design, you may only exclude those requirements that you do not perform. The organization must note which sections of 7.3 are excluded, as opposed to the entire 7.3 clause. For example, if your customer is responsible for product design but requires you to build a prototype and perform all or partial validation, then you may not exclude validation or prototype program. In the case of shared product design responsibility, the organization should clearly define its roles and responsibilities and how its processes link with its customer's processes. Also, keep in mind that the organization must conform to subclause 7.3.6.3 of the technical specification, which requires the product approval process recognized by the customer. I can think of no situations where this subclause could be excluded.

Another common problem third-party auditors see involves outsourced prototype processes. These may not be excluded and must be controlled by the organization (see Section 2.34). The organization may not delegate technical responsibility for outsourced processes.

Other areas organizations mistakenly try to exclude are process design, customer property, process validation, appearance items, and servicing. This is not permissible. Some of them, such as appearance items and servicing, may not be applicable at the time of the audit, but a system must be in place to handle them should they occur.

Another problem found during ISO/TS 16949:2002 audits is failure of the organization to justify the exclusions in the quality manual. You must document the justifications in the quality manual.

SECTION 2.5 UNAWARE OF CUSTOMER SCORECARDS

"What do you mean by customer scorecard?" Not what an auditor wants to hear!

Surprisingly, this is not such a rare occurrence. For example, it may be that the plant seeking certification has a corporate office that primarily handles customer requirements and orders and is listed as a remote location. The corporate office receives the plant's scorecards. Is this a problem? Yes, if the information is not shared with the plant so that the plant may address it. One major focus of the third-party audit is customer satisfaction. Customer scorecards, along with other information, will be used to develop the third-party audit plan, with the focus on weak areas. These scorecards must be available during both the Stage 1 Readiness Review and the Stage 2 Site Audit, and there should be evidence that the organization regularly reviews them and addresses weak areas.

SECTION 2.6 LINKAGES BETWEEN SITE AND REMOTE LOCATIONS NOT CLEARLY DEFINED

As we all know, the standard requires the processes and their interactions to be defined. All too often, organizations fail to do this for their remote locations, especially the interactions with the manufacturing site. This is extremely common in large corporations with corporate headquarters. Even if these headquarters were audited under QS-9000, their responsibilities were often not clearly defined. In fact, many times the manufacturing plant does not clearly understand these interactions and responsibilities. The new requirement will clearly help the organization to communicate to their people how these interactions work. And, by doing this in advance, it will help the organization to understand exactly which groups in their company need to be audited (see Chapter 3).

Consider a multisite company with one corporate person acting as the audit contact for all sites. The company discloses one corporate remote location during the Stage 1 Readiness Review. Is this complete? The third-party auditor will work to determine this. It may be possible for an individual manufacturing site to have other remote locations that the corporate contact is not aware of. As you can imagine, the Stage 2 audit would be a really bad time for the auditor to learn about the additional remote location. However, it would be a bad time for the corporate quality group to learn about it as well. It is important for the organization to use a multidisciplinary team to define its processes and how they interact, including responsibilities, to possibly avoid this situation.

Of course, in addition to assisting companies with defining their needed remote locations, these interactions and responsibilities must be documented to avoid nonconformities. Another benefit from defining these interactions is that you get an excellent tool for your internal auditors. Otherwise, without this roadmap, auditors would have difficulty providing an effective audit. Although written documents and tables may be adequate to do this, a detailed flowchart clearly indicating locations and responsibilities is highly recommended. A simplified example is shown in Figure 2.

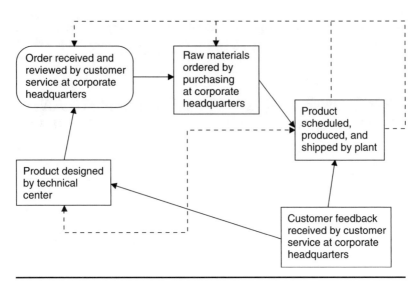

Figure 2   Linkages between site and remote locations.

SECTION 2.7 PROCESSES NOT FULLY DEFINED

As explained in Chapter 1, each process must be identified. These processes should include not only your manufacturing processes but the support processes as well. Typically, companies do rather well identifying their production, design, purchasing, and customer-related processes. The failure normally occurs with support processes such as training, internal quality audit, corrective and preventive action, and management review. These processes must be included.

There are many references out there covering definition and identification of processes as required by the ISO 9001:2000 standard and ISO/TS 16949:2002. I see identification of processes in a wide variety of ways, some very good and some very weak. The companies that seem to do the best job usually develop one top-level flowchart identifying all customer-oriented and support processes and their flow and linkages, as well as the inputs and outputs for each process. If the flowchart becomes too busy, some companies use a process matrix to accompany their process flow for the inputs and outputs. In addition, the process matrix may include the process owner and process goals. Often, a second tier of flowcharts is developed, one chart for each process identified on the top level.

After the processes are identified, the organization often fails to take the next step. The sequence and interaction of the identified processes must be defined. In other words, in what order do the processes occur? How do the processes work together? Typically, the simplest method to use is a process map that makes it easy to show the steps that occur and how they interact. For example, what comes first, an order or production? In an order-driven system, the order comes first. Prior to production, other processes such as purchasing occur. A process map clearly illustrates this sequence of events. Processes do not typically stand alone; instead, they interact with other processes. For example, purchasing must know what has been ordered by the customer and/or scheduled by production prior to ordering material from a supplier. As noted in Chapter 1, you should fully understand the interactions between the processes and work toward optimization of the system as a whole rather than optimization of each individual process.

SECTION 2.8 MEASURES NOT DEFINED FOR EVERY PROCESS

ISO/TS 16949:2002 clearly requires that the organization must monitor, measure, and analyze each of its processes. It often seems organizations begin the audit process assuming they have identified the metrics when they actually have not. Or, they actually have them but do not realize they have them.

First, let's talk about organizations that assume they have the measures they need. Typically, they have done very well with measures for manufacturing, because most companies have always done this. Metrics for the customer-related, purchasing, and design processes are often available but are usually somewhat weaker and not as well defined. Typically, the support processes create the problems, because these metrics are usually more difficult to define. And many organizations make the mistake of overlooking the measures for these processes. Examples of processes often overlooked would include training, internal quality audit, corrective action, and preventive action processes. Some examples of applicable measures for support processes are:

- Training: pretest score/posttest score, percent of employees completing annual training plans

- Internal quality audit: number of process audits completed on schedule, number of nonconformities issued (keep in mind that a higher number may not be bad), number of trained auditors

- Corrective action: number of customer complaints, number of corrective actions past due

Secondly, often the organization stumbles and simply cannot identify the process measures. More often than not it actually has these metrics but does not realize it. This results from lack of training and/or lack of understanding. Of course, as an auditor, I cannot give the answer even though I see it right in front of me! Be certain the leadership team has a strong understanding of process measures. They should understand their process measures better than their auditor, not the other way around!

I believe that the focus on measures is a great benefit in ISO/TS 16949:2002. Measures, if done correctly, drive the organization to meet common goals. On the other hand, if done haphazardly, they can be dangerous. People typically will focus their efforts to improve what the organization is monitoring. Thus, if the organization determines the wrong metrics, their personnel will be headed in the wrong direction!

Ensure the organization has a metric for each process identified, including support processes. Be safe. Why not include a column in your process matrix for measures? And be smart. The organization's leadership should spend time and take care developing the appropriate metrics to drive the organization in the right direction.

SECTION 2.9 LEGIBILITY OF DOCUMENTS NOT COVERED IN PROCEDURE

Although somewhat minor compared to other issues discussed thus far, almost every company initially excludes legibility of documents in its procedures. This occurs primarily because we are used to discussing legibility of records per ISO 1994/QS-9000, but legibility of documents is a new requirement. It expands the concept to include all documents, not just records. Because a record is a special type of document, if the organization's documented procedures just address records, this would not be adequate to meet the requirement. For example, if your organization allows personnel to handwrite changes or additions to procedures, it is logical that these handwritten changes and/or additions would be legible. Unfortunately, this is not always the case, as I can attest. Thus, the need to require legibility of documents is apparent.

Although this requirement may seem obvious to the organization, it must be documented.

SECTION 2.10 QUALITY OBJECTIVES NOT DEFINED AT RELEVANT FUNCTIONS AND LEVELS

Probably one of the weakest areas in many quality systems revolves around quality objectives. ISO/TS 16949:2002 tells us that they must be established at relevant functions and levels. What are relevant functions and levels? For optimal benefits, this would mean most functions and levels. Beginning at the top level of the organization, management should establish meaningful objectives. However, the organization should not stop there. Each area should establish objectives to support the top level. Then, these objectives should be carried through to the different levels. Objectives that are meaningful to top management may not be best for personnel in operations. Thus, it is important to put the objectives in terms that personnel can relate to. Everyone must be able to understand how they impact the objectives, as further discussed in Section 2.15.

Consider an organization with a quality objective to increase customer satisfaction. Each area should establish objectives to improve customer satisfaction to support the organization's goals. For example, purchasing may establish an objective to reduce material costs to partially pass on to the customer. Maintenance may establish an objective to reduce equipment breakdown, thus improving on-time delivery to the customer. This type of idea should be used for each process.

SECTION 2.11 RESPONSIBILITY FOR QUALITY NOT DEFINED AND/OR COMMUNICATED

The organization must define who is responsible for product quality, and that person(s) must be able to stop production to correct quality issues. Also, all shifts must have persons who are responsible for product quality and thus have the authority to stop production. This is often not clearly defined in organizations. It is not uncommon for production operators to tell me they are responsible for quality; however, when asked who can stop production, they many times refer to their supervisor. Occasionally on night shifts, I am told the person who can stop production is not there and must be called at home. These people should understand that this makes them responsible for quality on that shift. The organization should clearly define this and make certain every person understands the policy. The operator, supervisor, manager, and plant manager should all have the same understanding.

SECTION 2.12 COST OF POOR QUALITY NOT PROPERLY CALCULATED

The cost of quality was a measure used some time ago, and now it has resurfaced in ISO/TS 16949:2002. Although some companies have continued to use it over time, it is now a new measure for many organizations. The standard uses the cost of "poor" quality as opposed to the cost of quality, good and bad. Thus, at a minimum, the organization must know what the cost of poor quality is.

Not evaluating the cost of poor quality properly is a common error. Many companies, especially bulk suppliers, seem to use scrap only. This is not enough, because scrap is only part of it. The organization must consider other costs as well. Examples may include the cost to maintain the corrective action system, extra hours for sorting, travel costs and man-hours associated with customer complaints, and so on. Also, the measure must be a dollar value because we are dealing with cost.

The International Automotive Oversight Bureau (IAOB) does not allow a lenient interpretation of this measure. I found this out the hard way when I was written up during an IAOB witness audit. The client had only utilized scrap, and I had mistakenly accepted it. I will not make that mistake again.

SECTION 2.13 COMPETENCY NOT DEFINED

Section 3.9.12 of Q9000-2000 defines competence as "demonstrated ability to apply knowledge and skills."[7] How do you determine if your employees are competent? If someone was competent to do a job when hired 20 years ago, is that someone still competent today? These are questions the organization should consider. Just because an employee is highly educated does not mean that employee is competent for a certain position. The organization should determine what skills and education are needed for each position (i.e., job description). As individuals are hired, it should be determined whether or not they have the necessary skills and education. If a requirement is missing from the person's background, training requirements should be determined. However, this alone does not satisfy the competency requirement, which involves "demonstrated ability." Often this is satisfied through job qualification for hourly employees and performance reviews for salaried employees. Periodic performance evaluations may take care of the evaluation of ongoing competency.

The training process should be strongly linked to the achievement of continual improvement discussed in Section 2.45. Personnel should receive ongoing training to grow and discover new ideas to improve their processes.

SECTION 2.14 NO RECORD OF ON-THE-JOB TRAINING

ISO/TS 16949:2002 very clearly requires on-the-job training for people in new or changed positions that affect product quality. This also includes temporary and/or contract employees. Although this on-the-job training probably (hopefully) occurs, often a record is not maintained. In the past, this was especially true for temporary employees. Another problem area seems to occur with employees whose positions change. If a job is modified, the employee must be retrained on the job. Be certain to define and maintain a record of this training.

SECTION 2.15 CONTRIBUTION TO QUALITY OBJECTIVES NOT KNOWN

Just communicating the quality objectives is not enough to meet the requirements of ISO/TS 16949:2002. The organization must make certain that its employees know how they contribute to achievement of the quality objectives. Section 6.2.2.d of ISO/TS 16949:2002 is another area that gets written up quite often during an audit. Employees must understand more than just what the objectives are; they must understand how they contribute to achieving those objectives.

For example, if one of the organization's objectives is to improve on-time delivery, then all employees should understand their roles. Many times, personnel outside of production do not recognize this. For example, customer service representatives should understand that to improve on-time delivery, they must be certain the company is capable of meeting the customer's due date requirement prior to taking the order.

To further illustrate, consider a company with a quality objective to reduce the number of customer complaints. Occasionally, only the coordinator of the complaint system realizes the number of customer complaints. To meet the requirements of ISO/TS 16949:2002, the organization normally publishes these numbers. However, many individuals outside of production do not understand how they impact these objectives. These individuals might be in maintenance or purchasing. To improve, the company should ensure that these individuals understand how they can positively impact the number of customer complaints through areas such as on-time delivery (maintenance) and/or product quality (purchasing).

SECTION 2.16 NOTIFICATION OF CONSEQUENCES NOT DONE

Do your employees know how nonconformities will impact your customer? For example, if the outside diameter of part A is out of specification, what problems will this create for the customer? The organization must make employees aware of these issues. One way to do this is by using defective parts returned from the customer. Share them with your employees, and if possible, place them on display. Be certain to discuss why the defect was a problem for the customer. If the parts are not available, discuss the issue with the employees using photographs if possible. And don't forget to discuss potential problems. The employees should understand the consequences of shipping nonconforming parts and/or material.

SECTION 2.17 LACK OF PROCESS TO MOTIVATE

How does the organization motivate its employees to achieve quality objectives? The process to do this is often overlooked, and, subsequently, written up during TS audits. Many organizations incorrectly assume this process has to involve monetary rewards. Often, recognition is much more effective.

A common approach used is profit sharing or bonuses directly tied to the quality objectives. Another useful method sometimes utilized involves rewards and/or recognition for suggestions. Recognizing employees for ideas or a job well done is just as acceptable as paying them. Instead of money, some companies use items such as a hat, jacket, pen, and so on. Some companies simply verbally recognize the employees in front of their peers.

SECTION 2.18 LACK OF PROCESS TO MEASURE PERSON'S KNOWLEDGE OF JOB IMPORTANCE AND CONTRIBUTION TO QUALITY OBJECTIVES

In Section 2.15 we discussed the requirement that employees must understand how they contribute to the quality objectives. But ISO/TS 16949:2002 goes one step further. Not only do employees have to understand, but the organization must have a process to measure how well its employees understand. When asked about this process during an audit, organizations often do not have an answer because they have overlooked it. However, more times than not, they actually have the process and are not aware of it. Think about it. If you are conducting internal audits and asking your employees how they contribute to the quality objectives, you are actually gathering the data to measure how well they know. Also, the employees should be asked the importance of their jobs as well. If these questions are being asked during your internal audit, the only part of the process left to do is analyze the responses. If many employees are not aware of how they contribute to the objectives or of the importance of their job, an action plan should be developed to correct the situation. In other words, a corrective action should be issued.

SECTION 2.19 RISK ANALYSIS NOT DONE

Another requirement often overlooked involves risk analysis for new products. ISO/TS 16949:2002 requires the organization to determine the manufacturing feasibility of a proposed product during the contract review process.[8] This must be documented and it must include risk analysis. Normally, the problem here occurs with the documentation. I doubt many companies would actually operate without manufacturing feasibility analysis, but often there is no documentation to verify it occurred.

SECTION 2.20 MANUFACTURING PROCESS DESIGN EXCLUDED

The requirements for manufacturing process design seem to be some of the most difficult to meet for many organizations. These requirements may not be excluded, thus the organization must be able to show implementation. If a group in your organization at another site is responsible for your plant's new equipment, then that group must be audited as a remote location for your plant.

It is common for organizations to state that they have not had any new processing equipment in their plant for many years. If that is the case, the organization should consider using the scale up for a new part even if it was on older equipment. The organization should be able to show implementation of the requirements for manufacturing process design inputs and outputs by doing this. Even using preexisting equipment, the organization should consider inputs such as design data, targets for capability and cost, customer requirements, and lessons learned from previous projects. The manufacturing design outputs are also reasonable for preexisting equipment and include control plans, FMEAs, work instructions, etc.

SECTION 2.21 DESIGN AND DEVELOPMENT MEASUREMENTS NOT DEFINED

How well are your design and development projects doing? ISO/TS 16949:2002 requires that you define metrics for these projects at appropriate stages. In addition, data must be analyzed and reported during management review meetings. Examples of metrics include timing, costs, risks, and so on. Often these metrics are not defined; even more commonly, they are not summarized during management reviews.

SECTION 2.22 PRODUCT APPROVAL PROCESS NOT COMPLETE

In Section 7.3.6.3, ISO/TS 16949:2002 requires that the organization must follow a product approval process that its customer is familiar with. Typically, three different problems may occur with this requirement.

First, if the organization was certified to QS-9000, it often only used the Production Part Approval Process (PPAP).[9] Of course, this still meets the requirement for the U.S. auto companies, but it may not meet the requirements of your international customers. It is important to first determine the process your customer requires. The organization should not automatically assume that all of its customers require PPAP for their product approval process. And we should no longer use the acronym "PPAP" when referring to the product approval process, because many companies outside of the United States have their own processes.

Second, the organization may try to exclude all of 7.3, Design and Development, if it does not perform product design. As discussed earlier, the organization may not exclude manufacturing process design, and I have yet to see the product approval process correctly excluded.

A third problem sometimes encountered is that organizations previously certified to QS-9000 were familiar with the requirement for PPAP under 4.2, Quality System, in the QS standard. Now, the product approval process is found with design and development and is sometimes overlooked.

The customer's product approval process must be followed by the organization. If the customer has given you the authority to deviate, be certain to obtain a written waiver.

SECTION 2.23 SUPPLIER PRODUCT APPROVAL PROCESS NOT DONE

Over and over again I write a nonconformity against the requirement stating that the organization must pass along its customer's product approval process to its suppliers. Previously, the QS-9000 standard only required that the organization utilize some type of part approval process, but it did not necessarily have to be PPAP (except GM suppliers). This requirement is often overlooked because in ISO/TS 16949:2002 you will find it under Section 7.3.6.3 as part of design and development. It means that if you are required to PPAP for your customer, then you must require that your supplier PPAP for you. If you are required to follow another product approval process for your customer, then you must pass that requirement on to your supplier. If this is something your organization has not done in the past, it may be an extensive undertaking for you. This is especially true if your suppliers refuse to comply. As the organization prepares for ISO/TS 16949:2002, this requirement should be considered early in the process.

SECTION 2.24 REEVALUATION OF SUPPLIERS NOT DONE

"Reevaluation" of suppliers is another term used in ISO/TS 16949:2002 that we did not see in the QS-9000 standard. In the past, companies would select and evaluate their suppliers initially with little done afterward unless there was a problem. This is not acceptable with ISO/TS 16949:2002. The organization must reevaluate its suppliers at some point. It is recommended that this reevaluation be done at periodic intervals to assess quality, service, and delivery. Also, evaluate the supplier's response to any corrective action requests that may have been issued to it. Does the supplier create reoccurring problems for you? A summary of the supplier's performance should be included in management review.

Suppliers are so very critical to today's organizations, and their assessments can lead to improvement of your products. By tracking each supplier and assessing its performance, the organization is identifying problems and should be either forcing the supplier to improve or finding another supplier. A small investment in this area to develop the process may be very beneficial to the organization.

SECTION 2.25 ISO 9001:2000 CERTIFIED SUPPLIERS UNKNOWN

As part of the purchasing requirement, the organization must ensure that all of its suppliers are ISO 9001:2000 third-party registered. Although not in the standard, the IAOB tells us on its website that they believe the organization should be able to do this within the first cycle of its certificate.[10] This gives the organization three years from the time it is originally certified to ensure its suppliers are third-party registered. If the organization has suppliers that are not ISO certified at the time of its initial certification audit, there must be a plan in place to meet the three-year stipulation. Also, the organization should keep in mind that it must perform supplier development working toward conformity with ISO/TS 16949:2002. Ensuring its suppliers are ISO 9001:2000 registered is just the first step. Although there is no requirement that suppliers must become ISO/TS 16949:2002 registered, the organization must continue working with those that are only ISO 9001:2000 certified with the goal of conformity. If the organization chooses to require that its suppliers become ISO/TS 16949:2002 registered, it is the organization's choice to do so.

SECTION 2.26 PROMOTION OF SUPPLIER MANUFACTURING PERFORMANCE MONITORING NOT DONE

For whatever reason, the final requirement in purchasing is often overlooked. The technical specification requires that the organization encourage supplier monitoring of its manufacturing process performance. How can this be done? It is actually not difficult. It could easily be done with a statement to suppliers with another notification, such as a supplier performance scorecard. When the organization notifies its suppliers that they must be ISO 9001:2000 registered at a minimum, manufacturing process performance could be promoted.

SECTION 2.27 VALIDATION OF PROCESSES NOT DEFINED

One of the most common nonconformities during an ISO/TS 16949:2002 audit involves validation of processes. More often than not, organizations believe that if they have no "special" processes, this does not apply to them. This is a mistake. An addition in ISO/TS 16949:2002 that we did not see in QS-9000 requires validation to apply to every process. So, whether you have special processes or not, you must validate and revalidate each process. Be certain to define your process and identify the records.

SECTION 2.28 INCOMPLETE MEASUREMENT SYSTEM ANALYSIS

Although not a new requirement for automotive, a problem area continues to be variation studies for measuring instruments. ISO/TS 16949:2002 requires a study for each type of measuring device mentioned on the control plan. Typically, an auditor will sample control plans for measuring devices and ask to see the variation studies. Often, the studies are not available for each type of gage. Problems are found especially with instruments involving destructive testing or subjective analysis. Do not fail to conduct studies for all types of measuring devices.

SECTION 2.29 DEFINITION OF AUDIT CRITERIA, SCOPE, AND METHODS NOT GIVEN

For internal audits, ISO/TS 16949:2002 requires audit criteria, scope, frequency, and methods to be defined. Normally, frequency is not a problem, but the others are often not clear. ANSI/ISO/ASQ Q9000-2000 defines audit criteria as a "set of policies, procedures, or requirements used as a reference."[11] Thus, for your internal audit, you must define the documents to be used as a reference, such as procedures, ISO/TS 16949:2002 standard, customers' supplier quality manuals, and so on. The scope of the audit would address the extent of the audit, such as the scope on your ISO certificate. If all of the quality system was not going to be covered during the particular audit, the scope should state this. The methods to be used include your auditing techniques, such as document review, interview of personnel, review of records, and so on. Be certain these items are clearly defined for each audit.

SECTION 2.30 NO RECORD OF AUDIT OF EACH MANUFACTURING PROCESS

ISO/TS 16949:2002 very clearly tells us that the organization must conduct an audit of every manufacturing process. For example, if you have five manufacturing lines, you must audit each line. How often this is done is left up to the organization, but the timing must be effective. One method would include the use of the process control plan. Why not take the control plan and go through each step to ensure it is being done?

The failure here seems to fall on the record keeping. Organizations commonly fail to document the audit of each process. Of course, it would be practical to interview personnel working with each process, and thus it might be helpful to keep a sign-in sheet of personnel interviewed as a record to verify that each process was audited.

SECTION 2.31 INCOMPLETE AUDITOR QUALIFICATION

The organization must make certain it has internal auditors qualified to audit to ISO/TS 16949:2002. How do they become qualified? For the most part, that is up to the organization to define. If the organization wants to optimize its system, it will include in-depth training for the technical specification as well as the process approach. It is highly recommend that the organization has at least one auditor who has undergone an accredited lead auditor course.

Don't forget to review your customer-specific requirements for additional requirements that the customers might have.

SECTION 2.32 MATERIAL SHIPPED ON AUTHORIZATION IDENTIFICATION NOT DONE

Another requirement that is not new to automotive but is often overlooked involves nonconforming product and customer waivers. If the organization obtains a waiver from its customer to ship product that is out of specification, the organization must remember that material shipped on an authorization must be properly identified on each shipping container. This is a common problem especially for lower tier bulk suppliers. Be certain to define responsibility and the process for making sure this occurs. If a customer does not want the container identified, be certain to get this in writing from your customer.

SECTION 2.33 NEW CONTROL PLAN REQUIREMENTS NOT INCLUDED

Were you aware that Annex A of ISO/TS 16949:2002 requires the addition of two items to the control plan that were not expected in QS-9000? Most organizations are not and thus receive a nonconformity. The two new requirements are error proofing and corrective action. The new requirement for corrective action is in addition to reaction plan. It may be helpful for the organization to add columns to its control plans to address these two requirements.

SECTION 2.34 FAILURE TO IDENTIFY CONTROL OF OUTSOURCED PROCESSES

Per ISO N526R, the term "outsource" is interchangeable with "subcontract."[12] When the organization outsources a process that impacts product conformity, that process must be controlled. Such processes would include prototype programs as well as manufacturing. Thus, if you outsource these processes, you may not exclude them, and you must identify how you control them. Typically, this control would be through requirements addressed by ISO/TS 16949:2002 in purchasing.

It is a common mistake for the organization to omit the control of these subcontracted processes from its quality system. The system must include these processes in its scope, and the methods to manage and control these processes should be clearly defined. The organization should treat the suppliers of these processes as it does other suppliers to the organization. The suppliers should go through a selection and evaluation process, and periodic reevaluation should occur.

SECTION 2.35 FAILURE TO COMMUNICATE THE EFFECTIVENESS OF THE QUALITY MANAGEMENT SYSTEM

Often companies do not do an adequate job communicating the effectiveness of the quality management system. When I ask employees how their quality system is doing, the typical answer is something like, "pretty good." When I go on to ask how they know, it is not uncommon for them to tell me that management lets them know if they do something wrong. Thus, because they have not heard anything, everything must be okay. Certainly, that does not meet the intent of ISO/TS 16949:2002. It is important that the company communicates the effectiveness of the quality management system to all employees. It is recommended that this communication occur through more than one media.

First, the management should verbally discuss the effectiveness of the system. Let the employees know how well the company is doing. Tell them how well they are meeting their quality objectives. Let them know the level of customer satisfaction. If there have been customer complaints, share them with the employees. Occasionally, companies tell me they do not want to share the information because it is too confidential. This requirement does not mean that all metrics must be shared company-wide. However, measures should be developed and shared that measure the effectiveness of the quality system without compromising confidentiality of the company. Communication with your employees is instrumental in developing a strong quality system. They need to know where they stand and what needs improvement. Sometimes a company will bring all employees together annually to share the status of the business with them. This is a strong method for communication, but annually is not enough. Consider supplementing this type of annual meeting with more frequent but less extensive meetings. Perhaps the communication of objectives could be added to meetings that already exist, such as shift change meetings, weekly department meetings, safety meetings, and so on.

Second, it is highly recommended that the status of the objectives and other metrics be posted in various areas visible to the employees. When they are asked about the effectiveness of the quality system, they should have an answer and know what it is based on. When I am auditing and ask about the effectiveness of the system, I am highly encouraged when the employee takes me to charts of the metrics and explains them to me. Keep in mind that posting performance to objectives should not replace the verbal communication. Verbal communication is important to ensure the employee understands the meaning. Posting the objectives should be in support of the verbal explanation.

SECTION 2.36 INCOMPLETE MANAGEMENT REVIEW RECORDS

ISO/TS 16949:2002 is very specific about what is required for management review. More often than not when auditing companies new to ISO/TS 16949:2002, I issue a nonconformity against management review because the record is incomplete. It seems to be very difficult to remember to include all the requirements in the meeting. Companies that avoid this problem normally utilize a checklist as a supplement to their meeting minutes. It is recommended that you develop a checklist that includes all management review requirements of ISO/TS 16949:2002. An example is illustrated in Appendix H. It differs from Appendix E in that Appendix E is a checklist to be used during an internal audit; Appendix H is to be used during the management review itself. During your meeting, the checklist may then be used as an agenda and ist items checked off as they are discussed. Sometimes notes are taken directly on the checklist, but that is up to the company to determine. If you prefer to generate separate meeting minutes, that would suffice. However, one record may be easier to maintain.

SECTION 2.37 FAILURE TO EVALUATE EFFECTIVENESS OF TRAINING

Evaluating the effectiveness of training is not always a simple task; thus, often it is overlooked or ignored because the organization does not know how to perform the evaluation. There are several ways to evaluate this effectiveness, ranging from simple to elaborate methods. A few of the simpler methods are discussed in this section.

A common approach utilizes pretests and posttests. This way, you know how much subject knowledge the employees possessed coming in to the training and how much they gained during the training. With a little pre-planning, this method is simple to initiate and meets ISO/TS 16949:2002's requirements.

Another common approach is to evaluate the training during the employee's performance review. Did the training help the employee to improve job performance? This method should work, although the result may have been due to multiple factors, not just the training.

A third method sometimes used is to evaluate the organization or department's overall performance. Did the training provided improve the performance? This method is the least direct but is helpful in looking at the big picture.

I recommend utilizing a combination of all three methods to get the best evaluation of training. Training is normally a costly investment and should be monitored to ensure that the organization is getting its money's worth. For example, a large company may require each employee to obtain a significant amount of training hours each year, with accomplishment of this task linked to the bonus program. However, the subject of the training does not matter. The company should be commended for enriching its employee's lives, but it is suggested that at least some of the training be directly linked to the job for the organization to benefit as well as the individual. Evaluating training effectiveness should be a very helpful tool for the organization to learn which training helps it improve.

SECTION 2.38 METHOD OF CUSTOMER COMMUNICATION NOT ESTABLISHED

It is amazing how often I ask an organization about its process to communicate with customers and it seems surprised by the question. Perhaps this is because this requirement is much more clearly defined in ISO/TS 16949:2002 over the QS-9000 standard. ISO/TS 16949:2002 tells us that the organization must determine and implement effective arrangements to communicate with customers regarding product information, customer feedback (including complaints), questions, and orders (including amendments). How does the organization let the customer know who to contact for these issues? Often, this is handled by sales or customer service.

Or, perhaps it is because manufacturing plants sometimes assume customer communication is being covered by a corporate office. If this is the case, the process should be defined so that each location is aware of what is done. If the office that handles this is not part of the site being audited, it should be controlled as an outsourced process to ensure requirements of

ISO/TS 16949:2002 are being met (see Section 2.34). Conducting an internal audit of the office would be a method of evaluation and control.

SECTION 2.39 "REVALIDATION" OF PROCESSES NOT PERFORMED

Another new and often overlooked requirement involves the "revalidation" of all processes. Even companies that have "special" processes and thus were required to validate their processes under QS-9000 often miss this new requirement to revalidate in ISO/TS 16949. Validating the process at start-up is no longer enough. This is a requirement that some organizations already perform, but they do not seem to be ready for the question during an audit. Be prepared to define how you validate your processes as well as how you revalidate them. Depending upon the consequences of product failure, you should revalidate as often as needed for the process to be effective or as regulated by the government at a minimum for safety concerns.

SECTION 2.40 NO RECORD FOR CUSTOMER PROPERTY

This continues to be a problem during ISO/TS 16949:2002 audits just as it was with QS-9000 audits. Customer property involves anything that belongs to your customer that your organization controls. This may include raw material, labels, packaging, tooling, software, equipment, and so on. If anything of this nature is lost, damaged, or cannot be used for some reason, it must be reported to the customer, with a record maintained. It is not uncommon to find a weakness in this process. Often, the process is not well defined within the organization nor is the responsibility for the record. The organization should establish who is responsible for reporting the problem to the customer. In addition, the record should be established as a quality record. And if you have not yet experienced this problem, do not wait for the problem to occur to establish the record. Go ahead and define your process. You may have an empty file in the beginning, but you will be ready if the problem should occur.

Lost or damaged customer property typically does not occur on a regular basis, thus making the recording a process that is often difficult to remember when it does occur. The lack of keeping this record is a common nonconformity found during third-party audits. Perhaps it might be helpful to periodically review the process with the organization to ensure it is not forgotten when needed.

SECTION 2.41 RAW MATERIAL
SHELF LIFE UNKNOWN

While auditing preservation of product, I typically sample raw materials for indication of expiration date or shelf life. If any of the organization's raw materials have a shelf life, there should be a process to monitor and adhere to it. Though this process is normally found in organizations that have a large number of raw materials with shelf lives, it is often overlooked in organizations that have only a handful. Be certain the organization is aware of any raw materials that have an expiration date, and, if applicable, determine your process to handle them. Be cautious of certain resins, glues, paints, and so on.

SECTION 2.42 NO RECORD OF ASSESSMENT
FOR CALIBRATED EQUIPMENT
OUT OF TOLERANCE

What happens if you have a piece of equipment calibrated, and it is found out of tolerance? The common answer I receive is that the equipment is adjusted until it comes in tolerance. Yes, this correction should be made before the instrument is used again. However, this alone does not satisfy the requirements of ISO/TS 16949:2002. When an instrument is found outside of tolerance, ISO/TS 16949:2002 tells us that the validity of previous measurements must be assessed and recorded. In other words, you must assess the severity of the out-of-tolerance condition and determine if it impacted product measurements. If so, you must then assess the situation to try to determine when the out-of-tolerance condition developed, possibly going back to the prior calibration. The product measured with this instrument should be evaluated to determine if what the customer received was out-of-spec. A record for this entire assessment must be defined and maintained.

SECTION 2.43 NO PROACTIVE METHOD TO
DETERMINE CUSTOMER SATISFACTION

ISO/TS 16949:2002 clearly requires the organization to monitor customer satisfaction; however, it leaves it up to the organization to determine what and how. Often organizations, especially smaller ones, utilize customer complaints as their measure of customer satisfaction. Though this is indeed a measure, it should be coupled with a proactive approach such as a customer survey. This survey may be written or verbal, but it is important for the organization to seek out the information rather than wait for a complaint. If done properly, the proactive approach may prevent the problem from occurring or at least reduce its impact. If the decision is made to use a verbal survey, either formal or informal, it is important to record the results to assist with evaluation. And if you get a negative response, the organization should address it with the customer in a timely manner. If you are going to ask the customer how you are doing, you should be prepared to make the necessary adjustments to improve. If you do not address their concerns, the customers may stop responding, or worse yet, take their business elsewhere.

SECTION 2.44 NO EVIDENCE OF CONTINUAL IMPROVEMENT

I cannot tell you how many times I ask an organization for an example of continual improvement and it provides an example of a negative trend! Be prepared to discuss and provide examples of continual improvement in each process, including support. The continual improvement process may occur with breakthrough projects or through many small steps. Each area should be able to provide examples of projects for continual improvement as well as examples of positive trends. There is no requirement that the data be charted, but it certainly is easier for the organization to identify and prove positive trends with charts. I highly recommend the use of charts for data analysis. Not only does it make the audit easier but, more importantly, it helps the employees of your organization to visualize what is happening.

SECTION 2.45 DISREGARDING A REQUIREMENT

I have never understood why an organization simply ignores a requirement of ISO/TS 16949:2002. Perhaps it hopes the auditor will overlook the requirement or run out of time. This is a mistake for the organization. Address every requirement. It is far better to define your process to prompt thought and discussion for your third-party auditor rather than not address it at all. If you simply say you do not know or you forgot, the auditor will have no choice except to issue a nonconformance. For example, rather than ignoring training effectiveness because you do not feel you fully understand it, develop some type of process and implement it. Of course, you should do your best to obtain the information that you need through training, an expert in the field, and so on. The auditor will determine whether or not it meets the intent of ISO/TS 16949:2002, and at least you will have something in place for the requirement, which may possibly save you from getting a major nonconformity.

3

The Checklist

In chapters 1 and 2, this manual presented an overview of the process approach and a list of common mistakes encountered in ISO/TS 16949:2002 audits. Now it's time to move on to another topic: the process approach audit checklist, shown on the following pages in scaled-down size and included as Microsoft Word files in full-page size on the CD-ROM accompanying this book.

GUIDELINES FOR USE

All questions indicated with an asterisk in the second column of the checklist were derived from the ISO 9001:2000 standard (entry 3 in the list of references). All other questions were either paraphrased from ISO/TS 16949:2002 (entry 1 in the list of references) or contain requirements from both standards. The checklist is not meant to stand alone; it is important to use ISO/TS 16949:2002 along with the checklist to make certain the full meaning of the requirement is understood. The first step is to determine how the positions indicated on the checklist coincide with positions in your organization. For example, the checklist assumes you have both a purchasing manager and a shipping manager. If your organization combines these functions into a material manager, the auditor would not need to ask duplicate questions under purchasing and shipping when interviewing the material manager. Or, if you perform design but do not have a specific design lab, questions designated for the design lab tech should be merged in with the regular lab tech questions. If you find you do not have some positions indicated on the checklist, move the questions that have not been otherwise addressed to the appropriate area.

You may find that some of the questions are not in the appropriate process for your organization. For example, a question might be indicated for the quality manager when the engineering manager is actually responsible in your organization. When this is the case, the auditor should make note and pass the question on to the auditor responsible for auditing that area.

This checklist should be used as a tool and does not take the place of auditor training. It assumes the auditor will know what to do with the information after the question is asked; thus it should only be used by trained auditors. The auditors may gather some information themselves and may not need to ask every question. However, the auditor should keep in mind it is important not only to know that the requirements are met but also that the interviewee understands how. This is especially important for those preparing for a third-party audit. Also, the appendices are critical to the success of the audit and should be self-explanatory for a trained auditor.

Most importantly, for maximum benefits, the organization must add its own specific questions pertaining to procedures and work instructions. Also, don't forget to use Appendix F to develop questions from any customer-specific requirements that you may have.

Section 3.1 Management Process

	ISO/TS 16949:2002 Clause * indicates ISO 9001:2000	Function/Level	Question	Result
1	4.1.a*	Plant Manager	What process(es) do you manage and where are they identified?	
2	4.1.b*	Plant Manager	What is the order in which these processes occur and how do they interact?	
3	4.1.b*	Plant Manager	How do your processes link with the other processes in the company, outside of this plant?	

Auditor Name _____ Date of Audit _____

Section 3.1 Management Process

	ISO/TS 16949:2002 Clause * indicates ISO 9001:2000	Function/Level	Question	Result
4	4.1.d*/ 5.1.e*/ 6.1*	Plant Manager	How do you make certain resources and information needed to support the plant and monitoring of your processes are available?	
5	4.1.e*/ 8.2.3*	Plant Manager	How do you monitor and evaluate your processes? Please provide examples. For each process defined in #1, there should be a measurement.	
6	4.1*	Plant Manager	Do you outsource any processes? If so, how do you control these processes?	

Auditor Name _____ Date of Audit _____

Section 3.1 Management Process

	ISO/TS 16949:2002 Clause * indicates ISO 9001:2000	Function/Level	Question	Result
7	4.1.1	Plant Manager	*(Ask this question only if the answer to the previous question was yes.)* How do you make certain all customer requirements are met for these outsourced processes?	
8	4.2.3.1	Plant Manager	What is your process to review, hand out, and execute customer engineering standards, specifications, and changes?	
9	4.2.3.1	Plant Manager	How long does the review process just noted take? *(Must be done within two workweeks max.)*	
10	5.1.a*	Plant Manager	How do you convey the significance of meeting customer, regulatory, and legal requirements to your plant?	

Auditor Name _____   Date of Audit _____

Section 3.1 Management Process

	ISO/TS 16949:2002 Clause * indicates ISO 9001:2000	Function/Level	Question	Result
11	5.1.c*/ 5.4.1*	Plant Manager	What are the quality objectives for your plant and how were they developed?	
12	5.1.1	Plant Manager	How do you review your processes to ensure that they are successful and efficient? *(This would also include a review of the support processes.)*	
13	5.2*	Plant Manager	How do you make certain that customer requirements are determined and carried out with the goal of improving customer satisfaction?	

Auditor Name _____   Date of Audit _____

Section 3.1 Management Process

	ISO/TS 16949:2002 Clause * indicates ISO 9001:2000	Function/Level	Question	Result
14	5.3.e*	Plant Manager	ISO/TS 16949:2002 requires the quality policy to be reviewed for ongoing appropriateness. How is this done? *(Auditor should see some evidence that this has occurred.)*	
15	5.3.a*/ 5.1.b*	Plant Manager	How was your quality policy established? How is it suitable to the purpose of the company?	
16	5.3.b*	Plant Manager	*(The quality policy must have a vow to fulfill requirements and include continual improvement. If the auditor does not find this, ask the plant manager how this requirement is met.)*	

Auditor Name _____   Date of Audit _____

Section 3.1 Management Process

	ISO/TS 16949:2002 Clause * indicates ISO 9001:2000	Function/Level	Question	Result
17	5.3.c*	Plant Manager	How does the quality policy provide a structure for determining and evaluating quality objectives?	
18	5.3.d*	Plant Manager	How do you make certain the quality policy is conveyed and understood?	
19	5.4.1*	Plant Manager	For what functions and levels have quality objectives been developed and rolled out?	
20	5.4.1*	Plant Manager	Please explain how the objectives are quantifiable and constant with the policy.	

Auditor Name _____   Date of Audit _____

Section 3.1 Management Process

	ISO/TS 16949:2002 Clause * indicates ISO 9001:2000	Function/Level	Question	Result
21	5.4.1.1	Plant Manager	Show me where quality objectives and measures are incorporated in the business plan and how they are used to carry out the quality policy.	
22	5.4.1.1	Plant Manager	How do the quality objectives address the expectations of your customers?	
23	5.4.1.1	Plant Manager	Explain how your quality objectives are attainable within a definite time frame.	

Auditor Name _____   Date of Audit _____

Section 3.1 Management Process

	ISO/TS 16949:2002 Clause * indicates ISO 9001:2000	Function/Level	Question	Result
24	5.4.2.a*	Plant Manager	How is the planning and preparation of the quality system deployed to meet requirements and quality objectives?	
25	5.4.2.b*	Plant Manager	How do you make certain the integrity of the quality system is upheld when modifications are prepared and carried out? Please provide a specific example.	
26	5.5.1*	Plant Manager	Please show me where responsibilities and authorities are defined. How are they communicated?	

Auditor Name _____   Date of Audit _____

Section 3.1 Management Process

	ISO/TS 16949:2002 Clause * indicates ISO 9001:2000	Function/Level	Question	Result
27	5.5.1.1	Plant Manager	How do you make certain that appropriate managers are punctually notified when products or processes do not meet requirements?	
28	5.5.1.1	Plant Manager	Who has the right to stop production to rectify quality issues?	
29	5.5.1.1	Plant Manager	Who is responsible for product quality on each shift? *(Auditor should make certain this person(s) is the same as for the previous question.)*	
30	5.5.2*	Plant Manager	Who is your management representative, and how was this person appointed?	

Auditor Name _____   Date of Audit _____

Section 3.1 Management Process

	ISO/TS 16949:2002 Clause * indicates ISO 9001:2000	Function/Level	Question	Result
31	5.5.2*	Plant Manager	What are the management rep's responsibilities?	
31a	5.5.2.1	Plant Manager	Who has been assigned the responsibility to make sure all customer requirements are met?	
32	5.5.2.1	Plant Manager	Who is your customer representative assigned to make sure customer requirements are addressed in selecting special characteristics? In determining quality objectives? In corrective and preventive actions? And in product design and development?	
33	5.5.3*	Plant Manager	What communication processes are established within your plant?	

Auditor Name _____   Date of Audit _____

Section 3.1 Management Process

	ISO/TS 16949:2002 Clause * indicates ISO 9001:2000	Function/Level	Question	Result
34	5.5.3*	Plant Manager	How do you convey the success of your quality system to the plant?	
35	5.6*/ 4.2.4*/ 4.2.4.1	Plant Manager	*(Sample management review records to ensure they meet requirements of record control procedures. Use the Appendix C records form.)*	
36	5.6	Plant Manager	Describe your process for management review.	
37	5.6.1*	Plant Manager	How often do you conduct management reviews?	

Auditor Name _____   Date of Audit _____

Section 3.1 Management Process

	ISO/TS 16949:2002 Clause * indicates ISO 9001:2000	Function/Level	Question	Result
38	5.6	Plant Manager	May I see records of management reviews conducted since the last internal quality audit? *(Auditor to verify records show all requirements of ISO/TS 16949:2002 were met. Use Appendix E, Management Review.)*	
39	6.2.1*	Plant Manager	How do you determine competence of your employees?	
40	6.2.2.d*	Plant Manager	How do you make certain that employees in your plant are conscious of the importance of their jobs and how they add to the attainment of quality objectives?	

Auditor Name _____   Date of Audit _____

Section 3.1 Management Process

	ISO/TS 16949:2002 Clause * indicates ISO 9001:2000	Function/Level	Question	Result
41	6.2.2.4	Plant Manager	What is your process to measure the extent to which the employees are aware of the importance of their jobs and how they contribute to the quality objectives? Please show me the results.	
42	6.2.2.3	Plant Manager	How do you inform your employees about the penalty of nonconformities to your customers? Please provide specific examples.	
43	6.2.2.4	Plant Manager	What is your process to inspire personnel to meet quality objectives, continually improve, and construct an atmosphere to encourage innovation?	
44	6.2.2.4	Plant Manager	How do you encourage quality and technical awareness in your entire organization?	

Auditor Name _____   Date of Audit _____

Section 3.1 Management Process

	ISO/TS 16949:2002 Clause * indicates ISO 9001:2000	Function/Level	Question	Result
45	6.3.1	Plant Manager	How is the plant laid out to maximize effectiveness and efficiency?	
46	6.3.1	Plant Manager	How do you measure and examine the effectiveness of existing operations?	
47	6.3.2	Plant Manager	Show me your contingency plan to ensure customer requirements will be met in case of an emergency.	
48	6.4.1	Plant Manager	How does the plant tackle product safety and reduce possible risks to personnel, particularly in design and manufacturing?	

Auditor Name _____   Date of Audit _____

Section 3.1 Management Process

	ISO/TS 16949:2002 Clause * indicates ISO 9001:2000	Function/Level	Question	Result
49	7.1*	Plant Manager	What are your processes for product realization?	
50	7.1*	Plant Manager	What is the output of your planning of product realization?	
51	7.1.3	Plant Manager	How do you make certain privacy of customer-owned products and projects in process is maintained?	

Auditor Name _____   Date of Audit _____

Section 3.1 Management Process

	ISO/TS 16949:2002 Clause * indicates ISO 9001:2000	Function/Level	Question	Result
52	7.2.2.2	Plant Manager	When do you examine, verify, and record manufacturing viability of a projected product, including an analysis of risk? *(Auditor to verify answer when auditing appropriate process.)*	
53	7.4.1.1	Plant Manager	How do you make certain that procured materials used in products meet appropriate regulatory requirements?	
53a	7.5.1.7	Plant Manager	Describe the process to convey service issues to manufacturing, engineering, and design.	

Auditor Name _____   Date of Audit _____

Section 3.1 Management Process

	ISO/TS 16949:2002 Clause * indicates ISO 9001:2000	Function/Level	Question	Result
54	7.5.1.8	Plant Manager	Do you have service agreements with any of your customers? If so, please describe. *(Auditor to verify requirements are met with area responsible for service agreement.)*	
55	8.2.1*	Plant Manager	What is your process for acquiring and utilizing customer perception information?	
56	8.2.1.1	Plant Manager	Show me how you keep an eye on customer satisfaction.	

Auditor Name _____   Date of Audit _____

Section 3.1 Management Process

	ISO/TS 16949:2002 Clause * indicates ISO 9001:2000	Function/Level	Question	Result
56a	8.2.1.1	Plant Manager	Show me your performance measures, including quality performance of parts delivered, customer interruptions including field returns, on-time delivery, outbound premium freight occurrences, and customer notices pertaining to quality or delivery issues.	
57	8.2.1.1	Plant Manager	Show me how you keep an eye on the performance of the manufacturing processes to show fulfillment of customer requirements.	
58	8.2.3*	Plant Manager	What do you do when intended outcomes are not attained? *(Examples of correction and corrective action should be provided.)*	

Auditor Name _____   Date of Audit _____

Section 3.1 Management Process

	ISO/TS 16949:2002 Clause * indicates ISO 9001:2000	Function/Level	Question	Result
59	8.4*	Plant Manager	Please show me analysis of information pertaining to: 1. Customer satisfaction 2. Conformity of manufactured products 3. Characteristics and tendencies of processes and products, including prospects for preventive action, and 4. Vendors	
60	8.4.1	Plant Manager	Show me how you compare tendencies in performance with advancement toward objectives.	
61	8.4.1	Plant Manager	Show me how you use this data to weigh against rivals and/or benchmarks.	

Auditor Name _____   Date of Audit _____

Section 3.1 Management Process

	ISO/TS 16949:2002 Clause * indicates ISO 9001:2000	Function/Level	Question	Result
62	8.5.1.1	Plant Manager	What is your process for continual improvement?	
63	8.5.1*/ 8.5.1.2	Plant Manager	Please provide examples of continual improvement, including manufacturing process improvement.	
64	8.5.3*	Plant Manager	What is the difference between corrective action and preventive action?	
65	8.5.2/ 8.5.3*	Plant Manager	Have you personally participated in any corrective and/or preventive actions? If yes, please give an example.	

Auditor Name _____   Date of Audit _____

Section 3.1 Management Process

	ISO/TS 16949:2002 Clause * indicates ISO 9001:2000	Function/Level	Question	Result
66	4.2.2*	Management Representative	Show me how the quality manual meets requirements of the ISO/TS 16949:2002.	
67	6.1*	Determine by observation of plant overall	*(Do resources appear to be adequate to maintain the quality system and continually improve its effectiveness? ... to enhance customer satisfaction by meeting customer requirements?)*	
68	6.3*	Determine by observation of plant overall	*(Infrastructure: Does the plant include the appropriate workspace, equipment, and supporting services such as transport or communication?)*	
69	6.4*	Determine by observation of plant overall	*(Is the plant's work environment appropriate to meet product requirements?)*	

Auditor Name _____   Date of Audit _____

Section 3.1 Management Process

Use the following section to develop your own questions:

	Procedure/Work Instruction	Function/Level	Question	Result

Auditor Name _____   Date of Audit _____

Section 3.2 Customer-Related Process

	ISO/TS 16949:2002 Clause * indicates ISO 9001:2000	Function/Level	Question	Result
1	4.1.a*	Manager	What process(es) do you manage?	
2	4.1.b*	Manager	How does your process(es) link with the other processes in the company? *(Examples might be design, scheduling, production, etc.)*	
3	4.2.3.1	Manager	What is your group's role in the process to review, hand out, and execute customer engineering standards, specifications, and changes?	

Auditor Name _____   Date of Audit _____

Section 3.2 Customer-Related Process

	ISO/TS 16949:2002 Clause * indicates ISO 9001:2000	Function/Level	Question	Result
4	4.2.3.1	Manager	*(If this group is responsible for the process, verify that the review is done in no more than two workweeks. If it is not responsible, take this sample in the appropriate area.)* Customer Info Received Date Received Date Review Completed 1. 2. 3. 4. 5.	
5	5.1.a*	Manager	How do you communicate the importance of meeting customer, regulatory, and legal requirements to your area?	

Auditor Name _____   Date of Audit _____

Section 3.2 Customer-Related Process

	ISO/TS 16949:2002 Clause * indicates ISO 9001:2000	Function/Level	Question	Result
6	5.1.1	Manager	How do you assure your process(es) are successful and efficient?	
7	5.3.d*	Manager	How do you ensure the quality policy is communicated and understood?	
8	5.4.1*	Manager	What are the quality objectives for your area?	

Auditor Name _____   Date of Audit _____

Section 3.2 Customer-Related Process

	ISO/TS 16949:2002 Clause * indicates ISO 9001:2000	Function/Level	Question	Result
9	5.5.1*	Manager	Where are responsibilities and authorities defined, and how are they communicated?	
10	5.5.1*	Manager	How are duties divided up among the customer service reps (i.e., by region, by customer, etc.).	
11	5.5.1.1	Manager	How do you make certain that you are punctually notified when the customer-related process does not meet requirements?	

Auditor Name _____   Date of Audit _____

Section 3.2 Customer-Related Process

	ISO/TS 16949:2002 Clause * indicates ISO 9001:2000	Function/Level	Question	Result
12	5.5.3*	Manager	What communication processes are established within your area?	
13	5.5.3*	Manager	How do you communicate the effectiveness of your quality management system to your area?	
14	5.4.2*	Manager	How do you ensure the integrity of the quality management system is maintained when changes are planned and implemented?	

Auditor Name _____   Date of Audit _____

Section 3.2 Customer-Related Process

	ISO/TS 16949:2002 Clause * indicates ISO 9001:2000	Function/Level	Question	Result
15	5.6	Manager	What is your role in the management review meetings?	
16	6.2.1*	Manager	How do you determine competence of your employees?	
17	6.2.2.d*	Manager	How do you ensure that personnel in your area are aware of the importance of their jobs and how they contribute to achievement of quality objectives?	

Auditor Name _____   Date of Audit _____

Section 3.2 Customer-Related Process

	ISO/TS 16949:2002 Clause * indicates ISO 9001:2000	Function/Level	Question	Result
18	6.2.2.3	Manager	How do you inform your employees about the penalty of nonconformities pertaining to service to your customers?	
19	6.2.2.4	Manager	What is your process to inspire personnel to meet quality objectives, continually improve, and construct an atmosphere to encourage innovation?	
20	7.1*	Manager	What is your group's role in the planning of product realization?	

Auditor Name _____   Date of Audit _____

Section 3.2 Customer-Related Process

	ISO/TS 16949:2002 Clause * indicates ISO 9001:2000	Function/Level	Question	Result
21	7.2.1.a*/ Note 1	Manager	How do you determine requirements specified by the customer, including delivery and postdelivery requirements?	
22	7.2.1.b*/ Note 2	Manager	How do you determine requirements known for intended use even though not stated by the customer?	
23	7.2.1.c*/ Note 3	Manager	How do you determine regulatory and legal requirements of the product?	

Auditor Name _____   Date of Audit _____

Section 3.2 Customer-Related Process

	ISO/TS 16949:2002 Clause * indicates ISO 9001:2000	Function/Level	Question	Result
24	7.2.1.d*	Manager	How do you determine any other requirements?	
25	7.2.2*	Manager	What is your process to review the order prior to acceptance? *(Note: The answer should include ensuring product requirements are defined, resolving any differences, and ensuring the organization has the ability to meet the requirements.)*	
26	7.2.2*	Manager	Who has the authority to review the order to ensure all requirements can be met?	
27	7.2.2*	Manager	What is your record indicating the results of the review?	

Auditor Name _____   Date of Audit _____

Section 3.2 Customer-Related Process

	ISO/TS 16949:2002 Clause * indicates ISO 9001:2000	Function/Level	Question	Result
28	7.2.2*	Manager	What do you do if the customer does not provide documentation of its requirements? *(Note: The answer should include a method to confirm customer requirements prior to acceptance.)*	
29	7.2.2*	Manager	What happens if an order is changed? *(Note: Ensure that documents are updated and relevant personnel are notified of change.)*	
30	7.2.2.1	Manager	Do you ever waive the formal review of a contract/order? *(If yes, auditor must verify that customer authorization is available.)*	

Auditor Name _____   Date of Audit _____

Section 3.2 Customer-Related Process

	ISO/TS 16949:2002 Clause * indicates ISO 9001:2000	Function/Level	Question	Result
31	7.2.2.2	Manager	Show me how you examine, verify, and record manufacturing viability of a projected product during the contract review process, including risk analysis.	
32	7.2.3*	Manager	What is your process to communicate with your customer regarding product information, contracts, customer complaints, and so on?	
33	7.2.3.1	Manager	*(The auditor should determine what communication format the customers require. Ask the manager to verify that the organization has the ability to communicate in this format(s).)*	

Auditor Name _____   Date of Audit _____

Section 3.2 Customer-Related Process

	ISO/TS 16949:2002 Clause * indicates ISO 9001:2000	Function/Level	Question	Result
34	7.4.1.3	Manager	If your customers require that certain suppliers be utilized, how do you ensure this information is communicated to purchasing?	
34a	7.5.1.7	Manager	What is your process to communicate service concern information to manufacturing, design, and engineering?	
35	7.5.4*	Manager	Do you have any customer property? If so, how is it controlled? *(Note: This could include product, software, equipment, tooling, labels, etc.)*	
36	7.5.4*	Manager	What is your record for customer property that is lost, damaged, or otherwise unsuitable for use and to indicate it was reported to the customer?	

Auditor Name _____   Date of Audit _____

Section 3.2 Customer-Related Process

	ISO/TS 16949:2002 Clause * indicates ISO 9001:2000	Function/Level	Question	Result
37	8.2.1*	Manager	What are your customers' perceptions as to whether customer service has met customer requirements? How do you know this?	
37a	8.2.1*	Manager	Who are your internal customers and how do you measure their satisfaction?	
38	8.2.3*/ 4.1.e*	Manager	How do your monitor and measure your process? Please provide examples.	
39	8.2.3*	Manager	What do you do when planned results are not achieved? *(Examples of correction and corrective action should be provided.)*	

Auditor Name _____   Date of Audit _____

Section 3.2 Customer-Related Process

	ISO/TS 16949:2002 Clause * indicates ISO 9001:2000	Function/Level	Question	Result
40	8.4.a*	Manager	Show me analysis of customer satisfaction data.	
41	8.5.2	Manager	What internal corrective actions have been issued/assigned to or closed in your area(s) since the last internal audit? *(Auditor should randomly sample at least six corrective actions.)*	
42	8.5.2	Manager	What customer complaints have been issued/assigned to or closed in your area(s) since the last internal audit? *(Auditor should randomly sample at least six customer complaints.)*	

Auditor Name _____   Date of Audit _____

Section 3.2 Customer-Related Process

	ISO/TS 16949:2002 Clause * indicates ISO 9001:2000	Function/Level	Question	Result
43	8.5.3*	Manager	What preventive actions have been issued or closed in your area(s) since the last internal audit? *(Because this number is usually small, the auditor should review all preventive actions.)*	
44	8.5.1*	Manager	Please provide examples of continual improvement in your area(s) of responsibility.	

Auditor Name _____   Date of Audit _____

Section 3.2 Customer-Related Process

Interview as many customer service reps (or persons actually receiving the contract/order) as possible, asking each one of them questions 45–72 at a minimum:

	ISO/TS 16949:2002 Clause * indicates ISO 9001:2000	Function/Level	Question	Result
45	4.2.3*	All	*(Randomly sample at least eight documents in the area to ensure they meet requirements of document control procedures. Be sure to include the quality manual, procedures, work instructions, forms, and external documents. Use the Appendix B documents form.)*	
46	4.2.4*/ 4.2.4.1	Customer Service Rep	*(Randomly sample at least 12 records of contract review in the area to ensure they meet requirements of record control procedures. Use the Appendix C records form.)*	
47	5.1.a*	Customer Service Rep	How does the company communicate the importance of meeting customer, regulatory, and legal requirements to you?	

Auditor Name _____   Date of Audit _____

Section 3.2 Customer-Related Process

	ISO/TS 16949:2002 Clause * Indicates ISO 9001:2000	Function/Level	Question	Result
48	5.3.d*	Customer Service Rep	What does the quality policy mean to you?	
49	5.4.1*	Customer Service Rep	What are the quality objectives for your area?	
50	5.5.1*	Customer Service Rep	What are your primary responsibilities? *(If not documented, ensure this matches answer from manager.)*	
51	5.5.1*	Customer Service Rep	Whom do you work for? *(Ensure this matches answer from manager.)*	

Auditor Name _____   Date of Audit _____

Section 3.2 Customer-Related Process

	ISO/TS 16949:2002 Clause * Indicates ISO 9001:2000	Function/Level	Question	Result
52	5.5.3*	Customer Service Rep	How effective is your quality management system? How is this communicated to you?	
53	6.2.2.d*	Customer Service Rep	How do you contribute to the achievement of the quality objectives?	
54	6.2.2.d*	Customer Service Rep	What is the importance of your job?	
55	6.2.2.3	Customer Service Rep	When you first moved into this position, what type of training were you given? *(Auditor to verify on-the-job training was provided.)*	

Auditor Name _____   Date of Audit _____

Section 3.2 Customer-Related Process

	ISO/TS 16949:2002 Clause * indicates ISO 9001:2000	Function/Level	Question	Result
56	6.2.2.3	Customer Service Rep	If a service requirement is not met (such as date of delivery), what are the consequences to the customer?	
57	6.2.2.4	Customer Service Rep	Does the company have a process that inspires and encourages you to meet quality objectives, continually improve, and come up with new ideas? If so, what is this process?	
58	6.3*	Customer Service Rep	What additional equipment and tools do you need to do your job?	
59	7.2.1.a*/ Note 1	Customer Service Rep	How do you determine requirements specified by the customer, including delivery and postdelivery requirements?	

Auditor Name _____   Date of Audit _____

Section 3.2 Customer-Related Process

	ISO/TS 16949:2002 Clause * indicates ISO 9001:2000	Function/Level	Question	Result
60	7.2.1.b*/ Note 2	Customer Service Rep	How do you determine requirements known for intended use even though not stated by the customer?	
61	7.2.1.c*/ Note 3	Customer Service Rep	How do you determine regulatory and legal requirements of the product?	
62	7.2.1.d*	Customer Service Rep	How do you determine any other requirements?	
63	7.2.2*	Customer Service Rep	What is your process to review the order prior to acceptance? *(Note: The answer should include ensuring product requirements are defined, resolving any differences, and ensuring the organization has the ability to meet the requirements.)*	

Auditor Name _____   Date of Audit _____

Section 3.2 Customer-Related Process

	ISO/TS 16949:2002 Clause * Indicates ISO 9001:2000	Function/Level	Question	Result
64	7.2.2*	Customer Service Rep	Who has the authority to review the order to ensure all requirements can be met?	
65	7.2.2*	Customer Service Rep	What is your record indicating the results of the review?	
66	7.2.2*	Customer Service Rep	What do you do if the customer does not provide documentation of its requirements? *(Note: The answer should include a method to confirm customer requirements prior to acceptance.)*	
67	7.2.2*	Customer Service Rep	What happens if an order is changed? *(Note: Ensure that documents are updated and relevant personnel are notified of change.)*	

Auditor Name _____   Date of Audit _____

Section 3.2 Customer-Related Process

	ISO/TS 16949:2002 Clause * Indicates ISO 9001:2000	Function/Level	Question	Result
68	7.2.3*	Customer Service Rep	What is your process to communicate with your customer regarding product information, contracts, customer complaints, and so on?	
69	7.2.2.1	Customer Service Rep	Do you ever skip the requirement for a formal contract review? If so, who must authorize it?	
70	7.2.2.1	Customer Service Rep	For projected products, who is accountable for examining, verifying, and documenting their manufacturing viability, including risk analysis? Please explain.	

Auditor Name _____   Date of Audit _____

Section 3.2 Customer-Related Process

ISO/TS 16949:2002 Clause * indicates ISO 9001:2000		Function/Level	Question	Result
71	7.5.4*	Customer Service Rep	What is your record for customer property that is lost, damaged, or otherwise unsuitable for use and to indicate it was reported to the customer? *(Sample all records up to 12.)*	
72	8.5.2/ 8.5.3*	Customer Service Rep	Have you participated in any corrective and/or preventive actions? If yes, please give an example.	
73	6.1*	Determine by observation	*(Do resources appear to be adequate to maintain the quality system and continually improve its effectiveness? … to enhance customer satisfaction by meeting customer requirements?)*	
74	6.3*	Determine by observation	*(Infrastructure: Does the area include the appropriate workspace, equipment, and supporting services such as communication?)*	

Auditor Name _____   Date of Audit _____

Section 3.2 Customer-Related Process

ISO/TS 16949:2002 Clause * indicates ISO 9001:2000		Function/Level	Question	Result
75	5.1.a*	Supervisor	How do you communicate the importance of meeting customer, regulatory, and legal requirements to your area?	
76	5.3.d*	Supervisor	What does the quality policy mean to you? How do you ensure the quality policy is communicated and understood?	
77	5.4.1*	Supervisor	What are the quality objectives for your area?	

Auditor Name _____   Date of Audit _____

Section 3.2 Customer-Related Process

	ISO/TS 16949:2002 Clause * indicates ISO 9001:2000	Function/Level	Question	Result
78	5.5.3*	Supervisor	What communication processes do you use?	
79	5.5.3*	Supervisor	How effective is your quality management system? How do you know?	
80	6.2.2.d*	Supervisor	How do you contribute to the achievement of the quality objectives?	

Auditor Name _____   Date of Audit _____

Section 3.2 Customer-Related Process

	ISO/TS 16949:2002 Clause * indicates ISO 9001:2000	Function/Level	Question	Result
81	6.2.2.d*	Supervisor	What is the importance of your job?	
82	6.3*	Supervisor	Does your area need any equipment and/or tools?	
83	7.2.1.a*/ Note 1	Supervisor	How do you determine requirements specified by the customer, including delivery and postdelivery requirements?	

Auditor Name _____   Date of Audit _____

Section 3.2 Customer-Related Process

	ISO/TS 16949:2002 Clause * indicates ISO 9001:2000	Function/Level	Question	Result
84	7.2.1.b*/ Note 2	Supervisor	How do you determine requirements known for intended use even though not stated by the customer?	
85	7.2.1.c*/ Note 3	Supervisor	How do you determine regulatory and legal requirements of the product?	
86	7.2.1.d*	Supervisor	How do you determine any other requirements?	

Auditor Name _____   Date of Audit _____

Section 3.2 Customer-Related Process

	ISO/TS 16949:2002 Clause * indicates ISO 9001:2000	Function/Level	Question	Result
87	7.2.2*	Supervisor	What is your process to review the order prior to acceptance? *(Note: The answer should include ensuring product requirements are defined, resolving any differences, and ensuring the organization has the ability to meet the requirements.)*	
88	7.2.2*	Supervisor	Who has the authority to review the order to ensure all requirements can be met?	
89	7.2.2*	Supervisor	What is your record indicating the results of the review?	

Auditor Name _____   Date of Audit _____

Section 3.2 Customer-Related Process

	ISO/TS 16949:2002 Clause * indicates ISO 9001:2000	Function/Level	Question	Result
90	7.2.2*	Supervisor	What do you do if the customer does not provide documentation of its requirements? *(Note: The answer should include a method to confirm customer requirements prior to acceptance.)*	
91	7.2.2*	Supervisor	What happens if an order is changed? *(Note: Ensure that documents are updated and relevant personnel are notified of change.)*	

Auditor Name _____   Date of Audit _____

Section 3.2 Customer-Related Process

	ISO/TS 16949:2002 Clause * indicates ISO 9001:2000	Function/Level	Question	Result
92	7.2.2.1	Supervisor	Do you ever skip the requirement for a formal contract review? If so, who must authorize it?	
93	7.2.2.1	Supervisor	For projected products, who is accountable for examining, verifying, and documenting their manufacturing viability, including risk analysis? Please explain.	
94	7.2.3*	Supervisor	What is your process to communicate with your customer regarding product information, contracts, customer complaints, and so on?	

Auditor Name _____   Date of Audit _____

Section 3.2 Customer-Related Process

Use the following section to develop your own questions:

Procedure/Work Instruction	Function/Level	Question	Result

Auditor Name _____   Date of Audit _____

Section 3.3 Product Design Process

	ISO/TS 16949:2002 Clause * indicates ISO 9001:2000	Function/Level	Question	Result
1	7.3	Manager	(If 7.3, Product Design, was excluded from the scope, go through each section in 7.3 with the appropriate manager and determine whether or not it may be excluded at the current time. Use Section 3.3 of this checklist to audit subsections of the standard within scope.)	
2	4.2.3*	All	(Randomly sample at least 12 documents in the area to ensure they meet requirements of document control procedures. Be sure to include the quality manual, procedures, work instructions, forms, and external documents. Use the Appendix B documents form.)	
3	4.2.4*/ 4.2.4.1	All	(Randomly sample at least eight records in the area to ensure they meet requirements of record control procedures. Include records of design input, review, verification, validation, and changes. Use the Appendix C records form.)	
4	4.1.a*	Manager	What process(es) do you manage?	

Auditor Name _____   Date of Audit _____

Section 3.3 Product Design Process

	ISO/TS 16949:2002 Clause * indicates ISO 9001:2000	Function/Level	Question	Result
5	4.1.b*	Manager	How does your process(es) link with the other processes in the company?	
6	4.2.3.1	Manager	What is your group's role in the process to evaluate, hand out, and put into operation customer engineering standards, specifications, and changes?	
7	4.2.3.1	Manager	*(If this group is responsible for the process, verify that the review is done in no more than two workweeks.)* Customer Info Received Date Received Date Review Completed 1. 2. 3. 4. 5.	

Auditor Name _____   Date of Audit _____

Section 3.3 Product Design Process

	ISO/TS 16949:2002 Clause * indicates ISO 9001:2000	Function/Level	Question	Result
8	5.1.a*	Manager	How do you communicate the importance of meeting customer, regulatory, and legal requirements to your area?	
9	5.1.1	Manager	How do you make certain your process(es) are successful and capable?	
10	5.3.d*	Manager	How do you ensure the quality policy is communicated and understood?	
11	5.4.1*	Manager	What are the quality objectives for your area?	

Auditor Name _____   Date of Audit _____

Section 3.3 Product Design Process

	ISO/TS 16949:2002 Clause * indicates ISO 9001:2000	Function/Level	Question	Result
12	5.4.2*	Manager	How do you ensure the integrity of the quality management system is maintained when changes are planned and implemented?	
13	5.5.1*	Manager	Where are responsibilities and authorities defined, and how are they communicated?	
14	5.5.1.1	Manager	How do you ensure that you are punctually informed when products being designed and/or the design process do not conform to requirements?	
15	5.5.2.1	Manager	In product design and development, who represents the customers and makes certain their needs are met?	

Auditor Name _____   Date of Audit _____

Section 3.3 Product Design Process

	ISO/TS 16949:2002 Clause * indicates ISO 9001:2000	Function/Level	Question	Result
16	5.5.3*	Manager	What communication processes are established within your area?	
17	5.5.3*	Manager	How do you communicate the effectiveness of your quality management system to your area?	
18	6.2.1*	Manager	How do you determine competence of your employees?	
19	6.2.2*	Manager	What training is required for your employees?	
20	6.2.2.d*	Manager	How do you ensure that personnel in your area are aware of the importance of their job and how they contribute to achievement of quality objectives?	

Auditor Name _____   Date of Audit _____

Section 3.3 Product Design Process

	ISO/TS 16949:2002 Clause * indicates ISO 9001:2000	Function/Level	Question	Result
21	6.2.2.1	Manager	How do you make certain that persons working in your area are capable to attain design requirements?	
22	6.2.2.1	Manager	Show me where you have identified relevant tools and methods needed for design employees. How do you make certain persons working in your area are skilled in these relevant tools and methods?	
23	6.2.2.2	Manager	How and where do you identify training needs for persons performing design and development?	
24	6.2.2.2	Manager	How do you qualify design personnel?	
25	6.2.2.3	Manager	Please explain what on-the-the job training is conducted for your people.	

Auditor Name _____   Date of Audit _____

Section 3.3 Product Design Process

	ISO/TS 16949:2002 Clause * indicates ISO 9001:2000	Function/Level	Question	Result
26	6.2.2.3	Manager	Do you have any contract or agency personnel? If so, show me their on-the-job training record.	
27	6.2.2.3	Manager	How do you notify your people of the penalties to the customer when quality requirements are not achieved? Please provide specific examples.	
28	6.2.2.4	Manager	What is your process to inspire and encourage employees to meet quality objectives, continually improve, and promote new ideas?	
29	6.4.1	Manager	How do you speak to product safety and means to reduce possible risks to personnel in the design and development process?	

Auditor Name _____   Date of Audit _____

Section 3.3 Product Design Process

	ISO/TS 16949:2002 Clause * indicates ISO 9001:2000	Function/Level	Question	Result
30	7.1*	Manager	How does the design process interact with the planning of product realization as defined by the standard?	
30a	7.1/ Note	Manager	Do any of your customers require you to follow the APQP AIAG Reference Manual[13] or another product approval process? If so, please describe how this is done.	
30b	7.1*	Manager	During the planning process, how do you determine quality objectives and requirements for the product and the need to establish processes and documents and provide resources?	
31	7.1.1	Manager	How do you make sure customer requirements are included in the planning process as a part of the quality plan?	

Auditor Name _____   Date of Audit _____

Section 3.3 Product Design Process

	ISO/TS 16949:2002 Clause * indicates ISO 9001:2000	Function/Level	Question	Result
32	7.1.3	Manager	How does the company make certain that privacy is maintained for products and projects under development contracted by the customer?	
33	7.1.4	Manager	What is your process for product change control?	
34	7.1.4	Manager	How do you make sure product changes conform with customer requirements? How are these product changes validated?	
35	7.1.4	Manager	How do you make sure that impact on form, fit, and function for proprietary designs is reviewed with the customer? *(Sample and verify records to ensure customer reviews have occurred.)*	

Auditor Name _____   Date of Audit _____

Section 3.3 Product Design Process

	ISO/TS 16949:2002 Clause * indicates ISO 9001:2000	Function/Level	Question	Result
36	7.1.4	Manager	How do you make certain that further verification/identification requirements are met when required by the customer?	
37	7.2.2.2	Manager	What is your group's involvement in determining manufacturing viability of projected products during the contract review process, including risk analysis? *(Note: If design and development are the record keeper, sample and verify records.)*	
38	7.3.1*	Manager	How do you plan and control the design and development of product?	

Auditor Name _____   Date of Audit _____

Section 3.3 Product Design Process

	ISO/TS 16949:2002 Clause * indicates ISO 9001:2000	Function/Level	Question	Result
39	7.3.1*	Manager	How do you manage the interfaces between groups involved to ensure effective communication and clear assignment of responsibility?	
40	7.3.1.a*/ 7.3.1.b*	Manager	What are your design and development stages? For each stage, what review, verification, and validation are appropriate? Stage Review Verification Validation	

Auditor Name _____   Date of Audit _____

Section 3.3 Product Design Process

	ISO/TS 16949:2002 Clause * indicates ISO 9001:2000	Function/Level	Question	Result
41	7.3.1.c*	Manager	Please explain the responsibilities and authorities for design and development.	
42	7.3.1*	Manager	How do you make sure planning output is updated as the project progresses?	
43	7.3.1.1	Manager	To verify multidisciplinary approach, which functions of the organization participate in design and development?	
44	7.3.4* / Note	Manager	Explain how product design and manufacturing process design are coordinated and reviewed.	

Auditor Name _____   Date of Audit _____

Section 3.3 Product Design Process

	ISO/TS 16949:2002 Clause * indicates ISO 9001:2000	Function/Level	Question	Result
45	7.3.4.1	Manager	What measurements for product design and development have been defined and at what stages? *(Auditor to verify that these measurements have been analyzed and reported with summary results as an input to management review.)*	
45a	7.5.1.7	Manager	Describe the process to communicate service concerns to design.	
46	8.2.1*	Manager	What are your customers' perceptions as to whether design has met customer requirements? How do you know this?	

Auditor Name _____   Date of Audit _____

Section 3.3 Product Design Process

	ISO/TS 16949:2002 Clause * indicates ISO 9001:2000	Function/Level	Question	Result
46a	8.2.1*	Manager	Who are your internal customers and how do you measure their satisfaction?	
47	8.2.3*	Manager	What do you do when planned results are not achieved? *(Examples of correction and corrective action should be provided.)*	
48	8.2.3*/ 4.1.e*	Manager	How do you monitor and measure the design process? Please provide examples.	

Auditor Name _____ Date of Audit _____

Section 3.3 Product Design Process

	ISO/TS 16949:2002 Clause * indicates ISO 9001:2000	Function/Level	Question	Result
49	8.5.1*	Manager	Please provide examples of continual improvement in your area(s) of responsibility.	
50	8.5.2	Manager	What internal corrective actions have been issued/assigned to or closed in your area(s) since the last internal audit? *(Auditor should randomly sample at least six corrective actions.)*	
51	8.5.2	Manager	What customer complaints have been issued/assigned to or closed in your area(s) since the last internal audit? *(Auditor should randomly sample at least six customer complaints.)*	

Auditor Name _____ Date of Audit _____

Section 3.3 Product Design Process

	ISO/TS 16949:2002 Clause * indicates ISO 9001:2000	Function/Level	Question	Result
52	8.5.3*	Manager	What preventive actions have been issued or closed in your area(s) since the last internal audit? *(Because this number is usually small, the auditor should review all preventive actions.)*	
53	5.1.a*	Supervisor	How do you communicate the importance of meeting customer, regulatory, and legal requirements to your area?	
54	5.3.d*	Supervisor	What does the quality policy mean to you? How do you ensure the quality policy is communicated and understood?	

Auditor Name _____   Date of Audit _____

Section 3.3 Product Design Process

	ISO/TS 16949:2002 Clause * indicates ISO 9001:2000	Function/Level	Question	Result
55	5.4.1*	Supervisor	What are the quality objectives for your area?	
56	5.5.2.1	Supervisor	In product design and development, who represents the customers and makes certain their needs are met? *(Auditor to verify this matches answer from manager.)*	
57	5.5.3*	Supervisor	What communication processes do you use?	

Auditor Name _____   Date of Audit _____

Section 3.3 Product Design Process

	ISO/TS 16949:2002 Clause * indicates ISO 9001:2000	Function/Level	Question	Result
58	5.5.3*	Supervisor	How effective is your quality management system? How do you know?	
59	6.2.2.d*	Supervisor	How do you contribute to the achievement of the quality objectives?	
60	6.2.2.d*	Supervisor	What is the importance of your job?	
61	6.3*	Supervisor	What additional equipment and tools are needed in your area?	

Auditor Name _____   Date of Audit _____

Section 3.3 Product Design Process

	ISO/TS 16949:2002 Clause * indicates ISO 9001:2000	Function/Level	Question	Result
62	7.6.3.1	Supervisor	Do you have a design lab? *(If no, move to questions for design engineer. If yes, is it accredited to ISO/IEC 17025? If no, move to next question. If yes, view certificate to ensure the accreditation covers inspection and tests performed. If complete, skip the next two questions.)*	
63	7.6.3.1	Supervisor	Show me your design lab scope.	

Auditor Name _____   Date of Audit _____

Section 3.3 Product Design Process

	ISO/TS 16949:2002 Clause * indicates ISO 9001:2000	Function/Level	Question	Result
64	7.6.3.1	Supervisor	(Auditor to verify that the lab has specified and implemented: - Sufficient lab procedures - Capable design lab employees - Product testing - Ability to properly achieve these services, traceable to appropriate standard - Review of records)	
65	7.6.3.2/ Notes 1&2	Supervisor	Do you use any external labs for inspection, test, or calibration services pertaining to the design process? (If yes, verify that these labs are either acceptable to the customer, are the OEM, or are accredited.)	

Auditor Name _____   Date of Audit _____

Section 3.3 Product Design Process

Interview as many design engineers as possible, asking each one of them questions 66–105 at a minimum:

	ISO/TS 16949:2002 Clause * indicates ISO 9001:2000	Function/Level	Question	Result
66	5.1.a*	Design Engineer	How does the company communicate the importance of meeting customer, regulatory, and legal requirements to you?	
67	5.3.d*	Design Engineer	What does the quality policy mean to you?	
68	5.4.1*	Design Engineer	What are the quality objectives for your area?	

Auditor Name _____   Date of Audit _____

Section 3.3 Product Design Process

	ISO/TS 16949:2002 Clause * indicates ISO 9001:2000	Function/Level	Question	Result
69	5.5.1*	Design Engineer	What are your primary responsibilities? *(If not documented, ensure this matches answer from manager.)*	
70	5.5.1*	Design Engineer	Whom do you work for? *(Ensure this matches answer from manager.)*	
71	5.5.2.1	Design Engineer	In the product design and development process, who is the customer representative and what is that person's responsibility?	
72	5.5.3*	Design Engineer	How effective is your quality management system? How is this communicated to you?	

Auditor Name _____   Date of Audit _____

Section 3.3 Product Design Process

	ISO/TS 16949:2002 Clause * indicates ISO 9001:2000	Function/Level	Question	Result
73	6.2.2.d*	Design Engineer	How do you contribute to the achievement of the quality objectives?	
74	6.2.2.d*	Design Engineer	What is the importance of your job?	
75	6.2.2.1	Design Engineer	Show me where the organization has identified the tools and methods needed for the product design engineer to do his job. How are you qualified to use these tools and methods?	
76	6.2.2.3	Design Engineer	What on-the-job training did you receive?	

Auditor Name _____   Date of Audit _____

Section 3.3 Product Design Process

	ISO/TS 16949:2002 Clause * indicates ISO 9001:2000	Function/Level	Question	Result
77	6.2.2.4	Design Engineer	What process does the company use to inspire and encourage employees such as you to accomplish quality objectives, continually improve, and promote new ideas?	
78	6.3*	Design Engineer	What other equipment and tools do you need to do your job?	
79	6.4.1	Design Engineer	How does the company address product safety and ways to reduce possible risks to personnel in the design and development process?	

Auditor Name _____   Date of Audit _____

Section 3.3 Product Design Process

	ISO/TS 16949:2002 Clause * indicates ISO 9001:2000	Function/Level	Question	Result
80	7.3.2*/ Note	Design Engineer	What design projects have you recently completed? *(The auditor should randomly select at least one and ask to see records of design input. Ensure that input includes functional and performance requirements, applicable legal and regulatory requirements, information obtained from previous similar designs, and other necessary requirements.)*	
81	7.3.2*/ Note	Design Engineer	*(For each project selected, were inputs reviewed for adequacy?)*	

Auditor Name _____   Date of Audit _____

Section 3.3 Product Design Process

	ISO/TS 16949:2002 Clause * indicates ISO 9001:2000	Function/Level	Question	Result
82	7.3.2*/ Note	Design Engineer	*(For each project selected, were the requirements complete, clear, and not in conflict with one another?)*	
83	7.3.2.1	Design Engineer	*(For each project selected, ensure input requirements had been recognized, documented, and reviewed, including customer requirements, use of information, and goals.)*	
84	7.3.2.1	Design Engineer	What is your process for lessons learned from previous projects? *(Auditor to verify lessons learned had been utilized for projects sampled.)*	

Auditor Name _____   Date of Audit _____

Section 3.3 Product Design Process

	ISO/TS 16949:2002 Clause * indicates ISO 9001:2000	Function/Level	Question	Result
85	7.3.2.3/ Note	Design Engineer	How do you identify special characteristics? *(For projects sampled, ensure special characteristics were on the control plan, were in accordance with customer definition and symbols, and steps affecting them were included on process control documents such as operator instructions and FMEAs.)*	
86	7.3.3*	Design Engineer	For each project selected, what are the design outputs? *(Ensure these outputs meet input requirements, provide information for purchasing and production, contain or reference product acceptance criteria, and specify characteristics of the product essential for its safe and proper use.)*	
87	7.3.3*	Design Engineer	For each project selected, how were the outputs approved prior to release?	

Auditor Name _____   Date of Audit _____

Section 3.3 Product Design Process

	ISO/TS 16949:2002 Clause * indicates ISO 9001:2000	Function/Level	Question	Result
88	7.3.3.1	Design Engineer	*(For product design outputs, ensure each sampled project included:* - *DFMEA, outcome of reliability* - *Special characteristics and specs* - *Product error proofing as suitable* - *Product description with drawings or data* - *Results of product design reviews* - *Diagnostic procedures where appropriate)*	
89	7.3.4*	Design Engineer	When do design reviews occur?	
90	7.3.4*	Design Engineer	*(For each project selected, ask to see records of the results of the review and any necessary actions.)*	

Auditor Name _____   Date of Audit _____

Section 3.3 Product Design Process

	ISO/TS 16949:2002 Clause * indicates ISO 9001:2000	Function/Level	Question	Result
91	7.3.4*	Design Engineer	Who participated in these reviews? *(Ensure representatives of functions concerned were included.)*	
92	7.3.4.1	Design Engineer	How are projects measured?	
93	7.3.5*	Design Engineer	*(For each project selected, ask to see records of the results of verification. Ensure verification that design output met design input occurred.)*	

Auditor Name _____   Date of Audit _____

Section 3.3 Product Design Process

	ISO/TS 16949:2002 Clause * indicates ISO 9001:2000	Function/Level	Question	Result
94	7.3.6*/ Note 1	Design Engineer	*(For each project selected, ask to see records of validation results. Verify validation was done to ensure resulting product was capable of meeting requirements.)*	
95	7.3.6.1	Design Engineer	*(For each project selected, verify that validation was performed per customer requirements together with program timing.)*	
96	7.3.6*	Design Engineer	*(For each project selected, was validation completed prior to delivery or implementation of the product [if practical]?)*	

Auditor Name _____   Date of Audit _____

Section 3.3 Product Design Process

	ISO/TS 16949:2002 Clause * indicates ISO 9001:2000	Function/Level	Question	Result
97	7.3.7*/ Note	Design Engineer	*(For each project selected, ask to see records of design and development changes. Ensure each change was reviewed and approved prior to implementation.)*	
98	7.3.6.2	Design Engineer	*(For projects selected, did the customer require a prototype program and prototype control plan? If so, verify that it was completed. Verify that the same vendors, tooling, and manufacturing processes to be used when the product goes into production were utilized if possible.)*	
99	7.3.6.2	Design Engineer	How do you check performance testing activities for well-timed completion and conformance to requirements?	

Auditor Name _____   Date of Audit _____

Section 3.3 Product Design Process

	ISO/TS 16949:2002 Clause * indicates ISO 9001:2000	Function/Level	Question	Result
100	7.3.6.2	Design Engineer	Were prototype services outsourced? If so, how were they managed and controlled?	
101	7.3.6.3	Design Engineer	*(For each project sampled, what is the product approval procedure recognized by the customer? Verify the correct product approval procedure was followed.)*	

Auditor Name _____   Date of Audit _____

Section 3.3 Product Design Process

	ISO/TS 16949:2002 Clause * indicates ISO 9001:2000	Function/Level	Question	Result
102	7.3.6.3	Design Engineer	The product approval process accepted by the customer must be applied to your vendors. Does design or purchasing handle this? *(If design is responsible, auditor should sample records for materials to ensure the correct procedure was used. If purchasing is responsible, this information should be given to the auditor responsible for purchasing.)*	
103	7.3.7*/ Note	Design Engineer	How do you control design and development changes? *(For the projects sampled, ensure design and development changes were identified, with records maintained.)*	
104	7.3.7*/ Note	Design Engineer	How do you review, verify, validate, and approve the changes prior to implementation? *(For the projects sampled, be certain this was done for any changes that may have occurred.)*	

Auditor Name _____   Date of Audit _____

Section 3.3 Product Design Process

ISO/TS 16949:2002 Clause * indicates ISO 9001:2000	Function/Level	Question	Result
105	8.5.2/ 8.5.3*	Design Engineer	Have you participated in any corrective and/or preventive actions? If yes, please give an example.	
106	6.1*	Determine by observation in Design	*(Do resources appear to be adequate to maintain the quality system and continually improve its effectiveness? ... to enhance customer satisfaction by meeting customer requirements?)*	
107	6.3*	Determine by observation in Design	*(Infrastructure: Does the area include the appropriate workspace, equipment, and supporting services such as transport or communication?)*	

Auditor Name _____   Date of Audit _____

Section 3.3 Product Design Process

Interview as many design lab techs as possible, asking each one of them questions 108–127 at a minimum:

ISO/TS 16949:2002 Clause * indicates ISO 9001:2000	Function/Level	Question	Result
108	5.1.a*	Design Lab Tech	How does the company communicate the importance of meeting customer, regulatory, and legal requirements to you?	
109	5.4.1*	Design Lab Tech	What are the quality objectives for your area?	
110	5.5.1*	Design Lab Tech	What are your primary responsibilities? *(If not documented, ensure this matches answer from manager.)*	
111	5.5.1*	Design Lab Tech	Whom do you work for? *(Ensure this matches answer from manager.)*	

Auditor Name _____   Date of Audit _____

Section 3.3 Product Design Process

	ISO/TS 16949:2002 Clause * Indicates ISO 9001:2000	Function/Level	Question	Result
112	6.2.2.d*	Design Lab Tech	How do you contribute to the achievement of the quality objectives?	
113	6.2.2.d*	Design Lab Tech	What is the importance of your job?	
114	6.2.2.3	Design Lab Tech	What on-the-job training did you receive?	
115	6.2.2.3	Design Lab Tech	What current design project are you working on? If this finished product does not conform to quality requirements, what will be the penalty to the customer?	
116	6.2.2.4	Design Lab Tech	Does the company have a process to inspire and encourage employees to achieve quality objectives, and if so, what is it?	

Auditor Name _____   Date of Audit _____

Section 3.3 Product Design Process

	ISO/TS 16949:2002 Clause * Indicates ISO 9001:2000	Function/Level	Question	Result
117	6.3*	Design Lab Tech	What additional equipment and tools do you need to do your job?	
118	6.4.1	Design Lab Tech	During design and development, how has the company addressed product safety and ways to reduce possible risks to personnel?	
119	7.6*	Design Lab Tech	What monitoring and measuring devices do you use in the design lab? *(Sample devices using Appendix D, Calibration.)*	
120	7.6*	Design Lab Tech	How do you know these measuring devices are acceptable to use?	
121	7.6.3.1	Design Lab Tech	For the test that you are currently performing, show me the procedure.	

Auditor Name _____   Date of Audit _____

Section 3.3 Product Design Process

	ISO/TS 16949:2002 Clause * indicates ISO 9001:2000	Function/Level	Question	Result
122	8.2.4*	Design Lab Tech	How do you measure new designs to ensure requirements are met?	
123	8.2.4*	Design Lab Tech	What records do you keep for each project? *(Sample these records for projects sampled from design engineers. Ensure records demonstrate product/parts met requirements.)*	
124	8.3	Design Lab Tech	What do you do if a newly designed product or part does not meet requirements?	
125	8.5.2/ 8.5.3*	Design Lab Tech	Have you participated in any corrective and/or preventive actions? If yes, please give an example.	

Auditor Name _____   Date of Audit _____

Section 3.3 Product Design Process

	ISO/TS 16949:2002 Clause * indicates ISO 9001:2000	Function/Level	Question	Result
126	5.3.d*	Design Lab Tech	What does the quality policy mean to you?	
127	5.5.3*	Design Lab Tech	How effective is your quality management system? How is this communicated to you?	
128	6.1*	Determine by observation in Design Lab	*(Do resources appear to be adequate to maintain the quality system and continually improve its effectiveness? ... to enhance customer satisfaction by meeting customer requirements?)*	
129	6.3*	Determine by observation in Design Lab	*(Infrastructure: Does the area include the appropriate workspace, equipment, and supporting services such as transport or communication?)*	

Auditor Name _____   Date of Audit _____

Section 3.3 Product Design Process

Use the following section to develop your own questions:

Procedure/Work Instruction	Function/ Level	Question	Result

Auditor Name _____   Date of Audit _____

Section 3.4 Process Design

	ISO/TS 16949:2002 Clause * Indicates ISO 9001:2000	Function/Level	Question	Result
1	4.2.3*	All	*(Randomly sample at least six documents in the area to ensure they meet requirements of document control procedures. Be sure to include the quality manual, procedures, work instructions, forms, and external documents. Use the Appendix B documents form.)*	
2	4.2.4*/ 4.2.4.1	All	*(Randomly sample at least six records in the area to ensure they meet requirements of record control procedures. Include records of design input, review, verification, validation, and changes. Use the Appendix C records form.)*	
3	4.1.a*	Manager	What process(es) do you manage?	
4	4.1.b*	Manager	How does your process(es) link with the other processes in the company?	

Auditor Name _____   Date of Audit _____

Section 3.4 Process Design

	ISO/TS 16949:2002 Clause * indicates ISO 9001:2000	Function/Level	Question	Result
5	5.1.a*	Manager	How do you communicate the importance of meeting customer, regulatory, and legal requirements to your area?	
6	5.1.1	Manager	How do you make certain your process(es) are successful and capable?	
7	5.3.d*	Manager	How do you ensure the quality policy is communicated and understood?	
8	5.4.1*	Manager	What are the quality objectives for your area?	
9	5.4.2*	Manager	How do you ensure the integrity of the quality management system is maintained when changes are planned and implemented?	

Auditor Name _____   Date of Audit _____

Section 3.4 Process Design

	ISO/TS 16949:2002 Clause * indicates ISO 9001:2000	Function/Level	Question	Result
10	5.5.1*	Manager	Where are responsibilities and authorities defined, and how are they communicated?	
11	5.5.1.1	Manager	How do you ensure that you are punctually notified when processes being designed and/or the design process do not conform to requirements?	
12	5.5.3*	Manager	What communication processes are established within your area?	
13	5.5.3*	Manager	How do you communicate the effectiveness of your quality management system to your area?	

Auditor Name _____   Date of Audit _____

Section 3.4 Process Design

	ISO/TS 16949:2002 Clause * indicates ISO 9001:2000	Function/Level	Question	Result
14	6.2.1*	Manager	How do you determine competence of your employees?	
15	6.2.2*	Manager	What training is required for your employees?	
16	6.2.2.d*	Manager	How do you ensure that personnel in your area are aware of the importance of their jobs and how they contribute to achievement of quality objectives?	
17	6.2.2.2	Manager	How do you identify training needs for persons performing process design and development?	
18	6.2.2.2	Manager	How do you qualify process design employees?	

Auditor Name _____   Date of Audit _____

Section 3.4 Process Design

	ISO/TS 16949:2002 Clause * indicates ISO 9001:2000	Function/Level	Question	Result
19	6.2.2.3	Manager	Please explain what on-the-the job training is conducted for your people.	
20	6.2.2.3	Manager	Do you have any contract or agency personnel? If so, show me their on-the-job training records.	
21	6.2.2.4	Manager	What is your process to inspire and encourage employees to meet quality objectives, continually improve, and promote new ideas?	
22	6.3.1	Manager	How is a cross-functional approach used for developing plant, facility, and equipment plans?	

Auditor Name _____   Date of Audit _____

Section 3.4 Process Design

	ISO/TS 16949:2002 Clause * indicates ISO 9001:2000	Function/Level	Question	Result
23	6.3.1	Manager	When a new process design is begun, how do you make certain that the layout optimizes material movement, handling, and value-added utilization of the area?	
24	6.4.1	Manager	How do you address product safety and the ways to reduce possible risks to personnel in the design and development process?	
25	7.1*/ Note	Manager	How does the design process interact with the planning of product realization as defined by the standard?	
26	7.1.4/ Note 2	Manager	What is your process for manufacturing process change control? *(See Note 2 in 7.1.4 of the standard.)*	

Auditor Name _____   Date of Audit _____

Section 3.4 Process Design

	ISO/TS 16949:2002 Clause * indicates ISO 9001:2000	Function/Level	Question	Result
27	7.1.4	Manager	How do you make sure process changes are compliant with customer product requirements? How are these process changes validated?	
28	7.2.2.2	Manager	What is your group's involvement in determining manufacturing viability of planned products during the contract review process, including risk analysis? *(Note: If design and development are the record keeper, sample and verify records.)*	
29	7.3.1*	Manager	How do you plan and control the process design and development?	
30	7.3.1*	Manager	How do you manage the interfaces between groups involved to ensure effective communication and clear assignment of responsibility?	

Auditor Name _____   Date of Audit _____

Section 3.4 Process Design

	ISO/TS 16949:2002 Clause * indicates ISO 9001:2000	Function/Level	Question	Result
31	7.3.1.a*/ 7.3.1.b*	Manager	What are your design and development stages? For each stage, what review, verification, and validation are appropriate? Stage Review Verification Validation	
32	7.3.1.c*	Manager	Please explain the responsibilities and authorities for process design and development.	
33	7.3.1*	Manager	How do you make sure planning output is updated as the project progresses?	
34	7.3.1.1	Manager	To verify cross-functional approach, which functions of the organization participate in process design and development?	

Auditor Name _____   Date of Audit _____

Section 3.4 Process Design

	ISO/TS 16949:2002 Clause * indicates ISO 9001:2000	Function/Level	Question	Result
35	7.3.4*/ Note	Manager	Explain how product design and manufacturing process design are coordinated and reviewed.	
36	7.3.4.1/ Note	Manager	What metrics at specific stages of process design and development have been developed? *(Auditor to verify that these metrics have been examined and reported with summaries as an input to management review.)*	
36a	7.5.1.7	Manager	Describe the process to communicate service concerns to engineering.	
37	7.5.2*/ 7.5.2.1	Manager	How do you validate processes for production?	
37a	8.2.1*	Manager	Who are your internal customers and how do you measure their satisfaction?	

Auditor Name _____   Date of Audit _____

Section 3.4 Process Design

	ISO/TS 16949:2002 Clause * indicates ISO 9001:2000	Function/Level	Question	Result
38	8.2.3*	Manager	What do you do when planned results are not achieved? *(Examples of correction and corrective action should be provided.)*	
39	8.2.3*/ 4.1.e*	Manager	How do you monitor and measure the design process? Please provide examples.	
40	8.5.1*	Manager	Please provide examples of continual improvement in your area(s) of responsibility.	

Auditor Name _____   Date of Audit _____

Section 3.4 Process Design

	ISO/TS 16949:2002 Clause * indicates ISO 9001:2000	Function/Level	Question	Result
41	8.5.2	Manager	What internal corrective actions have been issued/assigned to or closed in your area(s) since the last internal audit? *(Auditor should randomly sample at least three corrective actions.)*	
42	8.5.2	Manager	What customer complaints have been issued/assigned to or closed in your area(s) since the last internal audit? *(Auditor should randomly sample at least four customer complaints.)*	
43	8.5.3*	Manager	What preventive actions have been issued or closed in your area(s) since the last internal audit? *(Because this number is usually small, the auditor should review all preventive actions.)*	

Auditor Name _____   Date of Audit _____

Section 3.4 Process Design

	ISO/TS 16949:2002 Clause * Indicates ISO 9001:2000	Function/Level	Question	Result
44	5.1.a*	Supervisor	How do you communicate the importance of meeting customer, regulatory, and legal requirements to your area?	
45	5.3.d*	Supervisor	What does the quality policy mean to you? How do you ensure the quality policy is communicated and understood?	
46	5.4.1*	Supervisor	What are the quality objectives for your area?	

Auditor Name _____   Date of Audit _____

Section 3.4 Process Design

	ISO/TS 16949:2002 Clause * Indicates ISO 9001:2000	Function/Level	Question	Result
47	5.5.3*	Supervisor	What communication processes do you use?	
48	5.5.3*	Supervisor	How effective is your quality management system? How do you know?	
49	6.2.2.d*	Supervisor	How do you contribute to the achievement of the quality objectives?	
50	6.2.2.d*	Supervisor	What is the importance of your job?	

Auditor Name _____   Date of Audit _____

Section 3.4 Process Design

	ISO/TS 16949:2002 Clause * indicates ISO 9001:2000	Function/Level	Question	Result
51	6.3*	Supervisor	Does your area need additional equipment and tools?	
52	7.5.2*/ 7.5.2.1	Supervisor	How do you validate processes for production?	

Interview as many process engineers as possible, asking each one of them questions 53–86 at a minimum:

	ISO/TS 16949:2002 Clause * indicates ISO 9001:2000	Function/Level	Question	Result
53	5.1.a*	Process Engineer	How does the company communicate the importance of meeting customer, regulatory, and legal requirements to you?	

Auditor Name _____   Date of Audit _____

Section 3.4 Process Design

	ISO/TS 16949:2002 Clause * indicates ISO 9001:2000	Function/Level	Question	Result
54	5.3.d*	Process Engineer	What does the quality policy mean to you?	
55	5.4.1*	Process Engineer	What are the quality objectives for your area?	
56	5.5.1*	Process Engineer	What are your primary responsibilities? *(If not documented, ensure this matches answer from manager.)*	
57	5.5.1*	Process Engineer	Whom do you work for? *(Ensure this matches answer from manager.)*	
58	5.5.3*	Process Engineer	How effective is your quality management system? How is this communicated to you?	

Auditor Name _____   Date of Audit _____

Section 3.4 Process Design

	ISO/TS 16949:2002 Clause * indicates ISO 9001:2000	Function/Level	Question	Result
59	6.2.2.d*	Process Engineer	How do you contribute to the achievement of the quality objectives?	
60	6.2.2.d*	Process Engineer	What is the importance of your job?	
61	6.2.2.3	Process Engineer	What on-the-job training did you receive?	
62	6.2.2.4	Process Engineer	What process does the company use to inspire and encourage employees to attain quality goals, continually improve, and promote new ideas?	
63	6.3*	Process Engineer	What additional equipment and tools do you need to do your job?	

Auditor Name _____   Date of Audit _____

Section 3.4 Process Design

	ISO/TS 16949:2002 Clause * indicates ISO 9001:2000	Function/Level	Question	Result
64	6.4.1	Process Engineer	How does the company address product safety and ways to reduce possible risks to employees in the design and development process?	
65	7.3.2*/ Note	Process Engineer	What design projects have you recently completed? *(The auditor should randomly select at least one and ask to see records of design input. Ensure that input includes functional and performance requirements, applicable legal and regulatory requirements, information obtained from previous similar designs, and other necessary requirements.)*	
66	7.3.2/ Note	Process Engineer	For the project selected, how were inputs reviewed for adequacy?	

Auditor Name _____   Date of Audit _____

Section 3.4 Process Design

	ISO/TS 16949:2002 Clause * indicates ISO 9001:2000	Function/Level	Question	Result
67	7.3.2/ Note	Process Engineer	For the project selected, describe how the requirements were complete, unambiguous, and in agreement with one another.	
68	7.3.2.2/ Note	Process Engineer	*(For the project selected, ensure manufacturing process design input requirements had been recognized, documented, and evaluated, including:* - *Product design output data* - *Targets for efficiency, process capability, and cost* - *Customer requirements, if any)*	
69	7.3.2.2/ Note	Process Engineer	What is your process for lessons learned from previous projects? *(Auditor to verify lessons learned had been utilized for project sampled.)*	

Auditor Name _____   Date of Audit _____

Section 3.4 Process Design

	ISO/TS 16949:2002 Clause * indicates ISO 9001:2000	Function/Level	Question	Result
70	7.3.2.3/ Note	Process Engineer	How do you identify special characteristics for process parameters? (See note 7.3.2.3.) *(For project sampled, ensure special characteristics were on the control plan, complied with customer definition and symbols, and steps affecting them were included on process control documents such as operator instructions and FMEAs.)*	
71	7.3.3*	Process Engineer	For the project selected, what are the design outputs? *(Ensure these outputs meet input requirements.)*	
72	7.3.3*	Process Engineer	For each project selected, how were the outputs approved prior to release?	

Auditor Name _____   Date of Audit _____

Section 3.4 Process Design

	ISO/TS 16949:2002 Clause * indicates ISO 9001:2000	Function/Level	Question	Result
73	7.3.3.2	Process Engineer	*(For manufacturing process design output, ensure the sampled project included:* - *Specifications and drawings* - *Manufacturing process flow chart/layout* - *PFMEAs* - *Control plan* - *Work instructions* - *Process approval acceptance criteria* - *Data for quality, reliability, maintainability, and measurability* - *Results of error-proofing activities, as appropriate* - *Methods of rapid detection and feedback of product/manufacturing process nonconformities)*	
74	7.3.4*/ Note	Process Engineer	When do design reviews occur?	
75	7.3.4*/ Note	Process Engineer	*(For each project selected, ask to see records of the results of the review and any necessary actions.)*	

Auditor Name _____   Date of Audit _____

Section 3.4 Process Design

	ISO/TS 16949:2002 Clause * indicates ISO 9001:2000	Function/Level	Question	Result
76	7.3.4*/ Note	Process Engineer	Who participated in these reviews? *(Ensure representatives of functions concerned were included.)*	
77	7.3.4.1/ Note	Process Engineer	How are projects measured?	
78	7.3.5*	Process Engineer	*(For each project selected, ask to see records of the results of verification. Ensure verification that design output met design input occurred. [See Note 2 under 7.3.6.])*	

Auditor Name _____   Date of Audit _____

Section 3.4 Process Design

	ISO/TS 16949:2002 Clause * indicates ISO 9001:2000	Function/Level	Question	Result
79	7.3.6*/ Note 2	Process Engineer	*(For each project selected, ask to see records of validation results. Verify validation was done to ensure resulting manufacturing process was capable of meeting requirements. [See Note 2 under 7.3.6 of the standard.])*	
80	7.5.2*/ 7.5.2.1	Process Engineer	How do you validate processes for production to demonstrate their ability to achieve planned results?	

Auditor Name _____   Date of Audit _____

Section 3.4 Process Design

	ISO/TS 16949:2002 Clause * indicates ISO 9001:2000	Function/Level	Question	Result
81	7.5.2*/ 7.5.2.1	Process Engineer	*(For the project selected, ask to see:* - *Defined criteria for review and approval of the process* - *Approval of equipment* - *Qualification of personnel* - *Specific methods and procedures* - *Requirements for requirements* - *Revalidation)* *(Note: if the process engineer is not responsible for all of these, determine who is and visit that person or pass information on to the appropriate auditor.)*	
82	7.3.7	Process Engineer	*(For the project selected, ask to see records of design and development changes. Ensure each change was reviewed and approved prior to implementation.)*	

Auditor Name _____   Date of Audit _____

Section 3.4 Process Design

	ISO/TS 16949:2002 Clause * indicates ISO 9001:2000	Function/Level	Question	Result
83	7.3.6.1	Process Engineer	For each project selected, show me how validation met customer requirements, including timing of program.	
84	7.3.7*/ Note	Process Engineer	How do you control design and development changes? *(For the project sampled, ensure design and development changes were identified, with records maintained.)*	
85	7.3.7*/ Note	Process Engineer	How do you review, verify, validate, and approve the changes prior to implementation? *(For the project sampled, be certain this was done for any changes that may have occurred.)*	
86	8.5.2/ 8.5.3*	Process Engineer	Have you participated in any corrective and/or preventive actions? If yes, please give an example.	

Auditor Name _____   Date of Audit _____

Section 3.4 Process Design

	ISO/TS 16949:2002 Clause * indicates ISO 9001:2000	Function/Level	Question	Result
87	6.1*	Determine by observation in Design	*(Do resources appear to be adequate to maintain the quality system and continually improve its effectiveness? ... to enhance customer satisfaction by meeting customer requirements?)*	
88	6.3*	Determine by observation in Design	*(Infrastructure: Does the area include the appropriate workspace, equipment, and supporting services such as transport or communication?)*	

Auditor Name _____   Date of Audit _____

Section 3.4 Process Design

Use the following section to develop your own questions:

Procedure/Work Instruction	Function/Level	Question	Result

Auditor Name _____   Date of Audit _____

Section 3.5 Purchasing Process

	ISO/TS 16949:2002 Clause * indicates ISO 9001:2000	Function/Level	Question	Result
1	4.2.3*	All	*(Randomly sample at least 12 documents in the area to ensure they meet requirements of document control procedures. Be sure to include the quality manual, procedures, work instructions, forms, raw material specifications, and external documents. Use the Appendix B documents form.)*	
2	4.2.4*/ 4.2.4.1	All	*(Randomly sample at least eight records in the area to ensure they meet requirements of record control procedures. Be sure to include purchase orders, raw material inspection records, certificates of analysis [if used], records of nonconforming raw materials/parts, and supplier evaluations. Use the Appendix C records form.)*	
3	4.1.a*	Manager	What process(es) do you manage?	

Auditor Name _____   Date of Audit _____

Section 3.5 Purchasing Process

	ISO/TS 16949:2002 Clause * indicates ISO 9001:2000	Function/Level	Question	Result
4	4.1.b*	Manager	How does your process(es) link with the other processes in the company?	
5	4.1*/ 4.1.1	Manager	Do you outsource any processes? If so, which ones and how do you control them?	
6	5.1.a*	Manager	How do you communicate the importance of meeting customer, regulatory, and legal requirements to your area?	

Auditor Name _____   Date of Audit _____

Section 3.5 Purchasing Process

	ISO/TS 16949:2002 Clause * indicates ISO 9001:2000	Function/Level	Question	Result
7	5.1.1	Manager	How do you make certain your process(es) are successful and capable?	
8	5.3.d*	Manager	How do you ensure the quality policy is communicated and understood?	
9	5.4.1*	Manager	What are the quality objectives for your area?	
10	5.4.2*	Manager	How do you ensure the integrity of the quality management system is maintained when changes are planned and implemented?	

Auditor Name _____   Date of Audit _____

Section 3.5 Purchasing Process

	ISO/TS 16949:2002 Clause * indicates ISO 9001:2000	Function/Level	Question	Result
11	5.5.1*	Manager	Where are responsibilities and authorities defined, and how are they communicated?	
12	5.5.1.1	Manager	If your process does not meet requirements, how do you know you will be punctually notified?	
13	5.5.3*	Manager	What communication processes are established within your area?	
14	5.5.3*	Manager	How do you communicate the effectiveness of your quality management system to your area?	

Auditor Name _____   Date of Audit _____

Section 3.5 Purchasing Process

	ISO/TS 16949:2002 Clause * indicates ISO 9001:2000	Function/Level	Question	Result
15	6.2.1*	Manager	How do you determine competence of your employees?	
16	6.2.2.d*	Manager	How do you ensure that personnel in your area are aware of the importance of their job and how they contribute to achievement of quality objectives?	
17	6.2.2.2	Manager	How do you qualify your personnel to do their job?	
18	6.2.2.3	Manager	What types of on-the-job training does your area provide?	
19	6.2.2.3	Manager	Do you have any contract or agency personnel? *(If so, auditor to ensure on-the-job training was provided.)*	

Auditor Name _____   Date of Audit _____

Section 3.5 Purchasing Process

	ISO/TS 16949:2002 Clause * indicates ISO 9001:2000	Function/Level	Question	Result
20	6.2.2.3	Manager	How do you inform your employees about the penalty of nonconformities to your customers? Please provide specific examples.	
21	6.2.2.4	Manager	What is your process to inspire and encourage employees to meet quality goals, to continually improve, and to promote new ideas?	
22	6.3.1	Manager	In the receiving area, how do you ensure the layout is optimized?	
23	6.4.1	Manager	In the receiving area, how do you address product safety and reduce possible risks to employees?	

Auditor Name _____   Date of Audit _____

Section 3.5 Purchasing Process

	ISO/TS 16949:2002 Clause * indicates ISO 9001:2000	Function/Level	Question	Result
24	6.4.2	Manager	In the receiving area, how do you address cleanliness?	
25	7.1	Manager	What is your role in the quality planning process?	
26	7.1.4	Manager	How do you address and control changes caused by a supplier?	
27	7.3.6.3	Manager	How do you make certain the product approval process required by your customer is applied to suppliers?	

Auditor Name _____   Date of Audit _____

Section 3.5 Purchasing Process

	ISO/TS 16949:2002 Clause * indicates ISO 9001:2000	Function/Level	Question	Result
28	7.4.1*	Manager	How do you make sure purchased product meets specified purchase requirements?	
29	7.4.1*	Manager	How do you control your raw material suppliers?	
30	7.4.1	Manager	How do you control other suppliers that impact your product and/or service such as carriers and providers of calibration services?	

Auditor Name _____   Date of Audit _____

Section 3.5 Purchasing Process

	ISO/TS 16949:2002 Clause * indicates ISO 9001:2000	Function/Level	Question	Result
31	7.4.1*	Manager	How do you select suppliers?	
32	7.4.1*	Manager	How do you evaluate and reevaluate suppliers?	
33	7.4.1/ Note	Manager	When mergers or acquisitions occur, how do you confirm the stability of the supplier's quality management system and its effectiveness?	

Auditor Name _____   Date of Audit _____

Section 3.5 Purchasing Process

	ISO/TS 16949:2002 Clause * Indicates ISO 9001:2000	Function/Level	Question	Result
34	7.4.1.1	Manager	How do you make certain that purchased materials or products meet appropriate regulatory requirements?	
35	7.4.1.2	Manager	Explain your process for supplier quality management system development.	
36	7.4.1.3	Manager	Do any of your customers specify from whom you must purchase? If so, what product and/or service and from whom?	

Auditor Name _____   Date of Audit _____

Section 3.5 Purchasing Process

	ISO/TS 16949:2002 Clause * Indicates ISO 9001:2000	Function/Level	Question	Result
37	7.4.2*	Manager	How do you ensure the purchase requirements are adequate prior to submittal to your supplier?	
38	7.4.3*	Manager	How do you ensure that purchased product meets your specified requirements?	
39	7.4.3*	Manager	Do you or your customer perform product verification at your supplier's facility? If yes, show me where the arrangements and method of product release are documented in the purchasing information.	

Auditor Name _____   Date of Audit _____

Section 3.5 Purchasing Process

	ISO/TS 16949:2002 Clause * indicates ISO 9001:2000	Function/Level	Question	Result
40	7.4.3.1	Manager	What method is used to ensure incoming product quality? *(Auditor to verify that one of the methods required in 7.4.3.1 is used.)*	
41	7.4.3.2	Manager	How do you monitor supplier performance? *(Auditor to verify that requirements of 7.4.3.2 are met.)*	
42	7.4.3.2	Manager	How do you encourage supplier monitoring of its manufacturing process performance?	

Auditor Name _____   Date of Audit _____

Section 3.5 Purchasing Process

	ISO/TS 16949:2002 Clause * indicates ISO 9001:2000	Function/Level	Question	Result
43	7.5.5*	Manager	Do you receive any purchased products or materials that need special handling/storage/protection? If so, please describe.	
44	7.5.5*	Manager	Do any of your purchased products or materials have a shelf life? If so, which ones and how do you manage them?	
45	7.5.5.1	Manager	How do you control out-of-date purchased product or material?	

Auditor Name _____   Date of Audit _____

Section 3.5 Purchasing Process

	ISO/TS 16949:2002 Clause * indicates ISO 9001:2000	Function/Level	Question	Result
46	7.5.5.1	Manager	Explain your process to appraise the state of purchased product in stock. How often is this done?	
47	7.5.5.1	Manager	What inventory management system is used to optimize purchased product inventory turns and guarantee stock rotation?	
47a	8.2.1*	Manager	Who are your internal customers and how do you measure their satisfaction?	

Auditor Name _____   Date of Audit _____

Section 3.5 Purchasing Process

	ISO/TS 16949:2002 Clause * indicates ISO 9001:2000	Function/Level	Question	Result
48	8.2.3*	Manager	What do you do when planned results are not achieved? *(Examples of correction and corrective action should be provided.)*	
49	8.2.3*/ 4.1.e*	Manager	How do you monitor and measure your process? Please provide examples.	
49a	8.3.4	Manager	If a supplier requests a waiver for product quality, what is the process to handle and notify the customer?	

Auditor Name _____   Date of Audit _____

Section 3.5 Purchasing Process

	ISO/TS 16949:2002 Clause * indicates ISO 9001:2000	Function/Level	Question	Result
50	8.4.d*	Manager	Show me analysis of supplier data.	
51	8.5.1*	Manager	Please provide examples of continual improvement in your area(s) of responsibility.	

Auditor Name _____   Date of Audit _____

Section 3.5 Purchasing Process

	ISO/TS 16949:2002 Clause * indicates ISO 9001:2000	Function/Level	Question	Result
52	8.5.2	Manager	What internal corrective actions have been issued/assigned to or closed in your area(s) since the last internal audit? *(Auditor should randomly sample at least six corrective actions.)*	
53	8.5.2	Manager	What customer complaints have been issued/assigned to or closed in your area(s) since the last internal audit? *(Auditor should randomly sample at least six customer complaints.)*	

Auditor Name _____   Date of Audit _____

Section 3.5 Purchasing Process

	ISO/TS 16949:2002 Clause * indicates ISO 9001:2000	Function/Level	Question	Result
54	8.5.3*	Manager	What preventive actions have been issued or closed in your area(s) since the last internal audit? *(Because this number is usually small, the auditor should review all preventive actions.)*	
55	7.4	While in Receiving	*(Sample raw materials/incoming parts using Appendix A, Purchasing. Complete the form as you go to purchasing and QC.)*	
56	5.1.a*	Receiving Supervisor	How do you communicate the importance of meeting customer, regulatory, and legal requirements to your area?	

Auditor Name _____   Date of Audit _____

Section 3.5 Purchasing Process

	ISO/TS 16949:2002 Clause * indicates ISO 9001:2000	Function/Level	Question	Result
57	5.3.d*	Receiving Supervisor	What does the quality policy mean to you? How do you ensure the quality policy is communicated and understood?	
58	5.4.1*	Receiving Supervisor	What are the quality objectives for your area?	
59	5.5.3*	Receiving Supervisor	What communication processes do you use?	
60	5.5.3*	Receiving Supervisor	How effective is your quality management system? How do you know?	

Auditor Name _____   Date of Audit _____

Section 3.5 Purchasing Process

	ISO/TS 16949:2002 Clause * indicates ISO 9001:2000	Function/Level	Question	Result
61	6.2.2.d*	Receiving Supervisor	How do you contribute to the achievement of the quality objectives?	
62	6.2.2.d*	Receiving Supervisor	What is the importance of your job?	
63	6.2.2.3	Receiving Supervisor	What type of training have you received?	
64	6.2.2.3	Receiving Supervisor	What on-the-job training is provided for your employees?	

Auditor Name _____   Date of Audit _____

Section 3.5 Purchasing Process

	ISO/TS 16949:2002 Clause * indicates ISO 9001:2000	Function/Level	Question	Result
65	6.2.2.4	Receiving Supervisor	What is the company's process to inspire and encourage employees to meet quality goals, to continually improve, and to promote new ideas?	
66	6.3*	Receiving Supervisor	What additional equipment and tools are needed in your area, if any?	
67	6.4*	Receiving Supervisor	How do you manage the work environment required to achieve conformity to product requirements?	
68	6.4.1	Receiving Supervisor	How does the organization address purchased product safety and the ways to reduce possible risks to personnel?	

Auditor Name _____   Date of Audit _____

Section 3.5 Purchasing Process

	ISO/TS 16949:2002 Clause * indicates ISO 9001:2000	Function/Level	Question	Result
69	7.4.1.1	Receiving Supervisor	How do you know that all purchased products and materials used in product meet appropriate regulatory requirements?	
70	7.5.1.1	Receiving Supervisor	Show me where receiving is included on the control plans.	
71	7.5.5*	Receiving Supervisor	Do you receive any purchased products or materials that need special handling/storage/protection? If so, please describe.	

Auditor Name _____   Date of Audit _____

Section 3.5 Purchasing Process

	ISO/TS 16949:2002 Clause * indicates ISO 9001:2000	Function/Level	Question	Result
72	7.5.5*	Receiving Supervisor	Do any of your purchased products or materials have a shelf life? If so, which ones and how do you manage them?	
73	7.5.5.1	Receiving Supervisor	How do you control outdated purchased product or material?	
74	7.5.5.1	Receiving Supervisor	Explain your process to evaluate the state of purchased product in stock. How often is this done?	
75	8.2.3*	Receiving Supervisor	How do you measure the receiving process?	

Auditor Name _____   Date of Audit _____

Section 3.5 Purchasing Process

	ISO/TS 16949:2002 Clause * indicates ISO 9001:2000	Function/Level	Question	Result
76	8.3	Receiving Supervisor	What is the process to handle nonconforming purchased product?	
77	8.3.1	Receiving Supervisor	What do you do with unidentified or suspect purchased product?	
78	8.5.2/ 8.5.3*	Receiving Supervisor	Describe any corrective and/or preventive actions that you have participated in.	
79	5.1.a*	Receiving Associate	How does the company communicate the importance of meeting customer, regulatory, and legal requirements to you?	

Auditor Name _____   Date of Audit _____

Section 3.5 Purchasing Process

	ISO/TS 16949:2002 Clause * indicates ISO 9001:2000	Function/Level	Question	Result
80	5.3.d*	Receiving Associate	What does the quality policy mean to you?	
81	5.4.1*	Receiving Associate	What are the quality objectives for your area?	
82	5.5.1*	Receiving Associate	What are your primary responsibilities? *(If not documented, ensure this matches answer from manager.)*	
83	5.5.1*	Receiving Associate	Whom do you work for? *(Ensure this matches answer from manager.)*	

Auditor Name _____   Date of Audit _____

Section 3.5 Purchasing Process

	ISO/TS 16949:2002 Clause * indicates ISO 9001:2000	Function/Level	Question	Result
84	5.5.3*	Receiving Associate	How effective is your quality management system? How is this communicated to you?	
85	6.2.2.d*	Receiving Associate	How do you contribute to the achievement of the quality objectives?	
86	6.2.2.d*	Receiving Associate	What is the importance of your job?	
87	6.2.2.2/ 6.2.2.3	Receiving Associate	What job training have you had?	

Auditor Name _____ Date of Audit _____

Section 3.5 Purchasing Process

	ISO/TS 16949:2002 Clause * indicates ISO 9001:2000	Function/Level	Question	Result
88	6.2.2.3	Receiving Associate	If a nonconforming purchased product or material ends up in your company's product and is shipped, what are the consequences to the customer?	
89	6.3*	Receiving Associate	What other equipment and tools do you need to do your job?	
90	6.4.1	Receiving Associate	How does the company address purchased product safety and ways to reduce possible risks to its employees?	
91	7.4.2*	Receiving Associate	What is your process to receive raw materials/parts?	

Auditor Name _____ Date of Audit _____

Section 3.5 Purchasing Process

	ISO/TS 16949:2002 Clause * indicates ISO 9001:2000	Function/Level	Question	Result
92	7.5.1.1/ 7.5.1.2	Receiving Associate	Show me your work instructions and control plan. *(Auditor to verify process used matches work instructions and control plan.)*	
93	7.5.3	Receiving Associate	How do you identify purchased product and materials?	
94	7.5.4*	Receiving Associate	How do you handle customer-owned material and/or parts that are received (if applicable)?	

Auditor Name _____   Date of Audit _____

Section 3.5 Purchasing Process

	ISO/TS 16949:2002 Clause * indicates ISO 9001:2000	Function/Level	Question	Result
95	7.5.5*	Receiving Associate	How do you handle and store raw materials/parts?	
96	7.5.5*	Receiving Associate	Do you receive any raw materials/parts that have special storage and/or protection needs? *(Examples would be temperature requirements such as refrigeration or heat, segregation for explosives, etc. Auditor to follow up with verification.)*	
97	7.5.5*	Receiving Associate	Do any of your purchased products or materials have a shelf life? If so, which ones and how do you handle them?	

Auditor Name _____   Date of Audit _____

Section 3.5 Purchasing Process

	ISO/TS 16949:2002 Clause * indicates ISO 9001:2000	Function/Level	Question	Result
98	7.5.5.1	Receiving Associate	What do you do with purchased product or material with an expired shelf life?	
99	7.6	Receiving Associate	What instruments do you use to weigh or otherwise measure purchased product or material, if any? *(Auditor to list instruments on Appendix D, Calibration.)*	
100	8.2.4	Receiving Associate	How do you measure incoming material/parts to ensure requirements are met?	
101	8.3	Receiving Associate	What do you do if incoming materials/parts do not meet requirements?	

Auditor Name _____   Date of Audit _____

Section 3.5 Purchasing Process

	ISO/TS 16949:2002 Clause * indicates ISO 9001:2000	Function/Level	Question	Result
102	8.3.1	Receiving Associate	What do you do with unidentified or suspect purchased product?	
103	8.5.2/ 8.5.3*	Receiving Associate	Describe any corrective and/or preventive actions that you have participated in.	
104	6.1*	Determine by observation in Receiving	*(Do resources appear to be adequate to maintain the quality system and continually improve its effectiveness? ... to enhance customer satisfaction by meeting customer requirements?)*	
105	6.3*	Determine by observation in Receiving	*(Infrastructure: Does the area include the appropriate workspace, equipment, and supporting services such as transport or communication?)*	

Auditor Name _____   Date of Audit _____

Section 3.5 Purchasing Process

	ISO/TS 16949:2002 Clause * indicates ISO 9001:2000	Function/Level	Question	Result
106	6.4*	Determine by observation in Receiving	*(Is the area's work environment appropriate to meet product requirements?)*	
107	6.4.2	Determine by observation in Receiving	*(Is the area in a state of order and clean?)*	
108	5.1.a*	Purchasing Supervisor	How do you communicate the importance of meeting customer, regulatory, and legal requirements to your area?	
109	5.3.d*	Purchasing Supervisor	What does the quality policy mean to you? How do you ensure the quality policy is communicated and understood?	

Auditor Name _____   Date of Audit _____

Section 3.5 Purchasing Process

	ISO/TS 16949:2002 Clause * indicates ISO 9001:2000	Function/Level	Question	Result
110	5.4.1*	Purchasing Supervisor	What are the quality objectives for your area?	
111	5.5.3*	Purchasing Supervisor	What communication processes do you use?	
112	5.5.3*	Purchasing Supervisor	How effective is your quality management system? How do you know?	
113	6.2.2.d*	Purchasing Supervisor	How do you contribute to the achievement of the quality objectives?	
114	6.2.2.d*	Purchasing Supervisor	What is the importance of your job?	

Auditor Name _____   Date of Audit _____

Section 3.5 Purchasing Process

	ISO/TS 16949:2002 Clause * indicates ISO 9001:2000	Function/Level	Question	Result
115	6.2.2.3	Purchasing Supervisor	What type of training have you received?	
116	6.2.2.3	Purchasing Supervisor	What on-the-job training is provided for your employees?	
117	6.2.2.4	Purchasing Supervisor	What is the company's process to inspire and encourage employees to meet quality goals, to continually improve, and to promote new ideas?	
118	6.3*	Purchasing Supervisor	What additional equipment, software, and so on is needed in your area, if any?	

Auditor Name _____   Date of Audit _____

Section 3.5 Purchasing Process

	ISO/TS 16949:2002 Clause * indicates ISO 9001:2000	Function/Level	Question	Result
119	6.4.1	Purchasing Supervisor	How does the organization address purchased product safety and the ways to reduce possible risks to personnel?	
120	7.4.1.1	Purchasing Supervisor	How do you know that all purchased products and materials used in product meet appropriate regulatory requirements?	
121	8.2.3*	Purchasing Supervisor	How do you measure the purchasing process?	
122	5.1.a*	Purchasing Associate/Buyer	How does the company communicate the importance of meeting customer, regulatory, and legal requirements to you?	

Auditor Name _____   Date of Audit _____

Section 3.5 Purchasing Process

	ISO/TS 16949:2002 Clause * indicates ISO 9001:2000	Function/Level	Question	Result
123	5.5.1*	Purchasing Associate/Buyer	What are your primary responsibilities? *(If not documented, ensure this matches answer from manager.)*	
124	5.5.1*	Purchasing Associate/Buyer	Whom do you work for? *(Ensure this matches answer from manager.)*	
125	5.3.d*	Purchasing Associate/Buyer	What does the quality policy mean to you?	
126	5.4.1*	Purchasing Associate/Buyer	What are the quality objectives for your area?	

Auditor Name _____   Date of Audit _____

Section 3.5 Purchasing Process

	ISO/TS 16949:2002 Clause * indicates ISO 9001:2000	Function/Level	Question	Result
127	5.5.3*	Purchasing Associate/Buyer	How effective is your quality management system? How is this communicated to you?	
128	6.2.2.d*	Purchasing Associate/Buyer	How do you contribute to the achievement of the quality objectives?	
129	6.2.2.d*	Purchasing Associate/Buyer	What is the importance of your job?	
130	6.2.2.3	Purchasing Associate/Buyer	What type of training have you received?	

Auditor Name _____   Date of Audit _____

Section 3.5 Purchasing Process

	ISO/TS 16949:2002 Clause * indicates ISO 9001:2000	Function/Level	Question	Result
131	6.2.2.4	Purchasing Associate/Buyer	What is the company's process to inspire and encourage employees to meet quality goals, to continually improve, and to promote new ideas?	
132	6.3*	Purchasing Associate/Buyer	What additional equipment, software, and so on do you need to do your job?	
133	7.4.1/ 7.4.2*	Purchasing Associate/Buyer	*(Continue samples from Appendix A, Purchasing.)*	
134	7.4.1.1	Purchasing Associate/Buyer	How do you know that all purchased products and materials used in product meets appropriate regulatory requirements?	

Auditor Name _____   Date of Audit _____

Section 3.5 Purchasing Process

	ISO/TS 16949:2002 Clause * indicates ISO 9001:2000	Function/Level	Question	Result
135	8.5.2/ 8.5.3*	Purchasing Associate/Buyer	Have you participated in any corrective and/or preventive actions? If yes, please give an example.	
136	6.1*	Determine by observation in Purchasing	*(Do resources appear to be adequate to maintain the quality system and continually improve its effectiveness? ... to enhance customer satisfaction by meeting customer requirements?)*	
137	6.3*	Determine by observation in Purchasing	*(Infrastructure: Does the area include the appropriate workspace, equipment, and supporting services such as transport or communication?)*	

Auditor Name _____   Date of Audit _____

Section 3.5 Purchasing Process

	ISO/TS 16949:2002 Clause * indicates ISO 9001:2000	Function/Level	Question	Result
138	5.1.a*	Incoming Lab/ QC Tech	How does the company communicate the importance of meeting customer, regulatory, and legal requirements to you?	
139	5.3.d*	Incoming Lab/ QC Tech	What does the quality policy mean to you?	
140	5.4.1*	Incoming Lab/ QC Tech	What are the quality objectives for your area?	

Auditor Name _____   Date of Audit _____

Section 3.5 Purchasing Process

	ISO/TS 16949:2002 Clause * indicates ISO 9001:2000	Function/Level	Question	Result
141	5.5.1*	Incoming Lab/ QC Tech	What are your primary responsibilities? *(If not documented, ensure this matches answer from manager.)*	
142	5.5.1*	Incoming Lab/ QC Tech	Whom do you work for? *(Ensure this matches answer from manager.)*	
143	5.5.3*	Incoming Lab/ QC Tech	How effective is your quality management system? How is this communicated to you?	
144	6.2.2.d*	Incoming Lab/ QC Tech	How do you contribute to the achievement of the quality objectives?	

Auditor Name _____   Date of Audit _____

Section 3.5 Purchasing Process

	ISO/TS 16949:2002 Clause * indicates ISO 9001:2000	Function/Level	Question	Result
145	6.2.2.d*	Incoming Lab/ QC Tech	What is the importance of your job?	
146	6.2.2.2/ 6.2.2.3	Incoming Lab/ QC Tech	What job training have you had?	
147	6.2.2.3	Incoming Lab/ QC Tech	If a nonconforming purchased product or material ends up in your company's product and is shipped, what are the consequences to the customer?	
148	6.3*	Incoming Lab/ QC Tech	What other equipment and tools do you need to do your job, if any?	
149	6.4.1	Incoming Lab/ QC Tech	How does the company address purchased product safety and ways to reduce possible risks to its personnel?	

Auditor Name _____   Date of Audit _____

Section 3.5 Purchasing Process

	ISO/TS 16949:2002 Clause * indicates ISO 9001:2000	Function/Level	Question	Result
150	7.1.2	Incoming Lab/ QC Tech	What is your acceptance criteria for attribute sampling, if applicable?	
151	7.4.1/ 7.4.2*/ 7.4.3*/ 7.4.3.1	Incoming Lab/ QC Tech	What is your process to assure the quality of purchased product? *(Auditor to continue completing Appendix A, Purchasing.)*	
152	7.5.1.1/ 7.6.3.1	Incoming Lab/ QC Tech	Show me your lab procedures and control plan. *(Auditor to verify process used matches work instructions and control plan.)*	
153	7.5.3	Incoming Lab/ QC Tech	How do you identify purchased product and materials?	

Auditor Name _____   Date of Audit _____

Section 3.5 Purchasing Process

	ISO/TS 16949:2002 Clause * indicates ISO 9001:2000	Function/Level	Question	Result
154	7.6	Incoming Lab/ QC Tech	What monitoring and measuring instruments do you use to test raw material/incoming parts? *(Sample devices using Appendix D, Calibration.)*	
155	7.6.2	Incoming Lab/ QC Tech	*(For each instrument sampled above, ensure records meet the requirements of 7.6.2.)*	
156	7.6	Incoming Lab/ QC Tech	How do you know these measuring devices are acceptable to use?	
157	7.6.3.1	Incoming Lab/ QC Tech	Show me your lab scope. *(Verify that it is a controlled document. Verify that all inspections, tests, and calibrations performed are on the lab scope.)*	

Auditor Name _____   Date of Audit _____

Section 3.5 Purchasing Process

	ISO/TS 16949:2002 Clause * indicates ISO 9001:2000	Function/Level	Question	Result
158	7.6.3.1	Incoming Lab/ QC Tech	How did you qualify for this position?	
159	7.6.3.2	Incoming Lab/ QC Tech	Is any purchased product sent to an external lab? *(If so, auditor to verify lab accreditation.)*	
160	8.2.4	Incoming Lab/ QC Tech	What records do you keep? *(Utilizing Appendix A, Purchasing, ensure records demonstrate product/parts met requirements and indicate who authorized released.)*	
161	8.3	Incoming Lab/ QC Tech	What do you do if incoming materials/parts do not meet requirements?	

Auditor Name _____   Date of Audit _____

Section 3.5 Purchasing Process

	ISO/TS 16949:2002 Clause * indicates ISO 9001:2000	Function/Level	Question	Result
162	8.5.2/ 8.5.3*	Incoming Lab/ QC Tech	Have you participated in any corrective and/or preventive actions? If yes, please give an example.	
163	6.1*	Determine by observation in Incoming Lab/QC	*(Do resources appear to be adequate to maintain the quality system and continually improve its effectiveness? ... to enhance customer satisfaction by meeting customer requirements?)*	
164	6.3*	Determine by observation in Incoming Lab/QC	*(Infrastructure: Does the area include the appropriate workspace, equipment, and supporting services such as transport or communication?)*	
165	6.4*	Determine by observation in Incoming Lab/QC	*(Is the area's work environment appropriate to meet product requirements?)*	

Auditor Name _____   Date of Audit _____

Section 3.5 Purchasing Process

Use the following section to develop your own questions:

Procedure/Work Instruction	Function/Level	Question	Result

Auditor Name _____   Date of Audit _____

Section 3.6 Production Process

	ISO/TS 16949:2002 Clause * indicates ISO 9001:2000	Function/Level	Question	Result
1	4.1.a*	Production Manager	What process(es) do you manage?	
2	4.1.b*	Production Manager	How does your process(es) link with the other processes in the company?	
3	4.1.1*	Production Manager	What production processes are outsourced, if any? *(Auditor to ensure these processes are controlled by the organization.)*	
4	4.2.3.1	Production Manager	Show me where you record the implementation date for changes implemented in production.	

Auditor Name _____   Date of Audit _____

Section 3.6 Production Process

	ISO/TS 16949:2002 Clause * indicates ISO 9001:2000	Function/Level	Question	Result
5	5.1.a*	Production Manager	How do you communicate the importance of meeting customer, regulatory, and legal requirements to your area?	
6	5.1.1	Production Manager	How do you make certain your process(es) are successful and capable?	
7	5.3.d*	Production Manager	How do you ensure the quality policy is communicated and understood?	
8	5.4.1*	Production Manager	What are the quality objectives for your area?	

Auditor Name _____   Date of Audit _____

Section 3.6 Production Process

	ISO/TS 16949:2002 Clause * indicates ISO 9001:2000	Function/Level	Question	Result
9	5.4.2*	Production Manager	How do you ensure the integrity of the quality management system is maintained when changes are planned and implemented?	
10	5.5.1*	Production Manager	Where are responsibilities and authorities defined, and how are they communicated?	
11	5.5.1.1	Production Manager	How do you make certain you are punctually informed when products or processes don't meet requirements?	
12	5.5.1.1	Production Manager	Who has the authority to stop production to fix quality concerns?	

Auditor Name _____   Date of Audit _____

Section 3.6 Production Process

	ISO/TS 16949:2002 Clause * indicates ISO 9001:2000	Function/Level	Question	Result
13	5.5.1.1	Production Manager	Who is responsible to ensure product quality on each shift? *(Note: Auditor to confirm that the person responsible to ensure product quality also has the authority to stop production to correct quality problems.)*	
14	5.5.3*	Production Manager	What communication processes are established within your area?	
15	5.5.3*	Production Manager	How do you communicate the effectiveness of your quality management system to your area?	
16	6.2.1*	Production Manager	How do you determine competence of your employees?	

Auditor Name _____   Date of Audit _____

Section 3.6 Production Process

	ISO/TS 16949:2002 Clause * indicates ISO 9001:2000	Function/Level	Question	Result
17	6.2.2.d*	Production Manager	How do you ensure that personnel in your area are aware of the importance of their job and how they contribute to achievement of quality objectives?	
18	6.2.2.2	Production Manager	How do you qualify employees?	
19	6.2.2.3	Production Manager	What on-the-job training is provided for your new employees? … for employees that have changed jobs? … for contract or agency personnel?	
20	6.2.2.3	Production Manager	How do you inform your employees about the penalty of nonconformities to your customers? Please provide specific examples.	

Auditor Name _____   Date of Audit _____

Section 3.6 Production Process

	ISO/TS 16949:2002 Clause * indicates ISO 9001:2000	Function/Level	Question	Result
21	6.2.2.4	Production Manager	What is your process to inspire and encourage employees to meet their quality goals, to make continual improvements, and to promote new ideas?	
22	6.3.1	Production Manager	Explain how the plant layout is optimized.	
23	6.3.1	Production Manager	How do you evaluate and monitor existing operations?	
24	6.3.2	Production Manager	Show me your contingency plan.	
25	6.4.1	Production Manager	How do you address product safety and the ways to reduce possible risks to employees?	

Auditor Name _____   Date of Audit _____

Section 3.6 Production Process

	ISO/TS 16949:2002 Clause * indicates ISO 9001:2000	Function/Level	Question	Result
26	6.4.2	Production Manager	How do you keep the production area orderly, clean, and in good state of repair?	
27	7.1*/ Note	Production Manager	How do you plan product realization?	
28	7.1.4/ Note 2	Production Manager	What is your process for process change control?	
29	7.1.4/ Note 2	Production Manager	How do you make sure process changes meet customer requirements? How are these process changes validated?	

Auditor Name _____   Date of Audit _____

Section 3.6 Production Process

	ISO/TS 16949:2002 Clause * indicates ISO 9001:2000	Function/Level	Question	Result
30	7.1.4/ Note 2	Production Manager	How do you make certain that additional verification/identification requirements are met when required by the customer?	
31	7.2.2*	Production Manager	When a customer order is amended, how is this communicated with production?	
32	7.2.2*	Production Manager	When customer requirements are changed, how is this communicated with production?	
33	7.2.2.2	Production Manager	How does the organization determine manufacturing viability of possible products?	

Auditor Name _____   Date of Audit _____

Section 3.6 Production Process

	ISO/TS 16949:2002 Clause * indicates ISO 9001:2000	Function/Level	Question	Result
34	7.3	Production Manager	How does production interact with the groups responsible for product and process design?	
35	7.3.1.1/ Note	Production Manager	What is your role in the design process?	
36	7.5.1*	Production Manager	How are production processes controlled?	
37	7.5.1.a*	Production Manager	Where would I find information about product characteristics?	

Auditor Name _____   Date of Audit _____

Section 3.6 Production Process

	ISO/TS 16949:2002 Clause * indicates ISO 9001:2000	Function/Level	Question	Result
38	7.5.1.b*/ 7.5.1.2	Production Manager	What format are production work instructions in and where would I find them?	
39	7.5.1.c*	Production Manager	How do you ensure your area is using suitable equipment?	
40	7.5.1.1/ Annex A	Production Manager	Show me your control plans. *(Auditor to verify requirements in Annex A of the standard are met.)*	
41	7.5.1.3	Production Manager	When and how are job setups verified?	
42	7.5.1.4	Production Manager	Describe your preventive and predictive maintenance process.	

Auditor Name _____   Date of Audit _____

Section 3.6 Production Process

	ISO/TS 16949:2002 Clause * indicates ISO 9001:2000	Function/Level	Question	Result
43	7.5.1.5	Production Manager	How is production tooling managed?	
44	7.5.1.6	Production Manager	How do you make certain production is scheduled to meet customer requirements?	
45	7.5.1.7	Production Manager	Describe the process to communicate service concerns to manufacturing.	
46	7.5.2*/ 7.5.2.1	Production Manager	How do you validate and revalidate processes per 7.5.2 of ISO/TS 16949:2002?	

Auditor Name _____   Date of Audit _____

Section 3.6 Production Process

	ISO/TS 16949:2002 Clause * indicates ISO 9001:2000	Function/Level	Question	Result
47	7.5.3*/ Note	Production Manager	How is product identified in your areas?	
48	7.5.3*/ 7.5.3.1	Production Manager	How is traceability maintained and what is the record of unique product identification?	

Auditor Name _____   Date of Audit _____

Section 3.6 Production Process

	ISO/TS 16949:2002 Clause * indicates ISO 9001:2000	Function/Level	Question	Result
49	7.5.3	Production Manager	*(The auditor should randomly sample a minimum of six products that shipped since the last internal audit. Trace each of these products back to the raw materials/incoming parts used. Ask the production manager to provide someone to assist you when you finish with his/her interview.)* 1. 2. 3. 4. 5. 6.	
50	7.5.4*	Production Manager	Do you have any customer property? If yes, how is it controlled? If it is lost, damaged, or otherwise unsuitable for use, how is this reported to the customer? What is the record?	

Auditor Name _____   Date of Audit _____

Section 3.6 Production Process

	ISO/TS 16949:2002 Clause * indicates ISO 9001:2000	Function/Level	Question	Result
51	7.5.4.1	Production Manager	How is the ownership of production tooling known?	
52	7.5.5*	Production Manager	How is product preserved during production and delivery?	
53	7.5.5*	Production Manager	What controls do you have in place for proper handling, packaging, storage, and protection of product?	
54	7.5.5.1	Production Manager	What are your intervals to review the state of product in stock and notice deterioration?	
55	7.5.5.1	Production Manager	Describe your inventory management system.	

Auditor Name _____   Date of Audit _____

Section 3.6 Production Process

	ISO/TS 16949:2002 Clause * indicates ISO 9001:2000	Function/Level	Question	Result
56	7.5.5.1	Production Manager	How do you handle outdated product?	
57	8.1.2	Production Manager	How do you ensure basic statistical concepts are understood and used?	
58	8.2.1*/ 8.4	Production Manager	What are your customers' perceptions as to whether your product has met their requirements? What data does your area collect and analyze pertaining to customer satisfaction?	
59	8.2.1.1	Production Manager	How do you examine the performance of manufacturing processes to show fulfillment with customer requirements for product quality and competence of process?	

Auditor Name _____   Date of Audit _____

Section 3.6 Production Process

	ISO/TS 16949:2002 Clause * indicates ISO 9001:2000	Function/Level	Question	Result
60	8.2.3*	Production Manager	What do you do when planned results are not achieved? *(Examples of correction and corrective action should be provided.)*	
61	8.2.3*/ 4.1.e*	Production Manager	How do you monitor and measure your process? Please provide examples. This should include data pertaining to characteristics and trends of process.	
62	8.2.3.1	Production Manager	How do you monitor and measure manufacturing processes?	

Auditor Name _____   Date of Audit _____

Section 3.6 Production Process

	ISO/TS 16949:2002 Clause * indicates ISO 9001:2000	Function/Level	Question	Result
63	8.2.3.1	Production Manager	How do you make certain that the capability or performance specified by the customer part approval process is sustained? *(Auditor should sample parts with special characteristics to ensure capability is monitored and maintained.)*	
64	8.2.3.1	Production Manager	Where are important process events such as tool change or machine repair documented? *(Auditor to sample and verify.)*	
65	8.2.3.1	Production Manager	When a manufacturing process becomes unstable or is no longer capable, how do you make sure a corrective action is completed? Please show me examples.	
66	8.2.3.1	Production Manager	Show me your records for process change effective dates.	

Auditor Name _____   Date of Audit _____

Section 3.6 Production Process

	ISO/TS 16949:2002 Clause * indicates ISO 9001:2000	Function/Level	Question	Result
67	8.2.4*/ Note	Production Manager	How do you monitor and measure your product? What records are kept as evidence of conformity? What records are kept to indicate who authorized release? This should include data pertaining to characteristics and trends of product.	
68	8.2.4.2	Production Manager	What parts do you produce that are designated as appearance items by your customers?	
69	8.3	Production Manager	What is your process to control nonconforming product?	
70	8.3.1	Production Manager	How do you classify unidentified or suspect product?	
71	8.3.2	Production Manager	Show me your instructions for rework.	

Auditor Name _____   Date of Audit _____

Section 3.6 Production Process

	ISO/TS 16949:2002 Clause * indicates ISO 9001:2000	Function/Level	Question	Result
72	8.3.3	Production Manager	How do you make certain the customer is punctually informed if you find nonconforming product has been shipped?	
73	8.3.4	Production Manager	Explain your process for customer waivers.	
74	8.4.c*	Production Manager	What data does your area collect and analyze pertaining to opportunities for preventive actions?	
75	8.5.1*	Production Manager	Please provide examples of continual improvement in your area(s) of responsibility.	

Auditor Name _____   Date of Audit _____

Section 3.6 Production Process

	ISO/TS 16949:2002 Clause * indicates ISO 9001:2000	Function/Level	Question	Result
76	8.5.1.2/ Note 1/ Note 2	Production Manager	Show me how manufacturing process improvement continually focuses on control and reduction of variation in product characteristics and process parameters.	
77	8.5.2	Production Manager	What internal corrective actions have been issued/assigned to or closed in your area(s) since the last internal audit? *(Auditor should randomly sample at least six corrective actions.)*	
78	8.5.2	Production Manager	What customer complaints have been issued/assigned to or closed in your area(s) since the last internal audit? *(Auditor should randomly sample at least six customer complaints.)*	

Auditor Name _____   Date of Audit _____

Section 3.6 Production Process

	ISO/TS 16949:2002 Clause * indicates ISO 9001:2000	Function/Level	Question	Result
79	8.5.2.2	Production Manager	Show me examples of where error proofing was used in the corrective action process.	
80	8.5.2.2	Production Manager	Show me examples of where controls to eliminate the causes of problems were applied to similar processes and products.	
81	8.5.3*	Production Manager	What preventive actions have been issued or closed in your area(s) since the last internal audit? *(Because this number is usually small, the auditor should review all preventive actions.)*	
82	5.1.a*	Production Supervisor/ Team Leader	How do you communicate the importance of meeting customer, regulatory, and legal requirements to your area?	

Auditor Name _____   Date of Audit _____

Section 3.6 Production Process

	ISO/TS 16949:2002 Clause * indicates ISO 9001:2000	Function/Level	Question	Result
83	5.3.d*	Production Supervisor/ Team Leader	What does the quality policy mean to you? How do you ensure the quality policy is communicated and understood?	
84	5.4.1*	Production Supervisor/ Team Leader	What are the quality objectives for your area?	
85	5.5.1.1	Production Supervisor/ Team Leader	Who has the ability to stop production to fix quality issues?	
86	5.5.1.1	Production Supervisor/ Team Leader	Who is responsible to ensure product quality on your shift? *(Note: Auditor to confirm that the person responsible to ensure product quality also has the authority to stop production to correct quality problems.)*	

Auditor Name _____   Date of Audit _____

Section 3.6 Production Process

	ISO/TS 16949:2002 Clause * indicates ISO 9001:2000	Function/Level	Question	Result
87	5.5.3*	Production Supervisor/ Team Leader	What communication processes do you use?	
88	5.5.3*	Production Supervisor/ Team Leader	How effective is your quality management system? How do you know?	
89	6.2.2.d*	Production Supervisor/ Team Leader	How do you contribute to the achievement of the quality objectives?	
90	6.2.2.d*	Production Supervisor/ Team Leader	What is the importance of your job?	

Auditor Name _____   Date of Audit _____

Section 3.6 Production Process

	ISO/TS 16949:2002 Clause * indicates ISO 9001:2000	Function/Level	Question	Result
91	6.2.2.2	Production Supervisor/ Team Leader	How do you qualify employees?	
92	6.2.2.3	Production Supervisor/ Team Leader	What on-the-job training is provided for your new employees? … for employees that have changed jobs? … for contract or agency personnel?	
93	6.2.2.3	Production Supervisor/ Team Leader	How do you inform your employees about the penalty of nonconformities to your customers? Please provide specific examples.	

Auditor Name _____   Date of Audit _____

Section 3.6 Production Process

	ISO/TS 16949:2002 Clause * indicates ISO 9001:2000	Function/Level	Question	Result
94	6.2.2.4	Production Supervisor/ Team Leader	What is your process to inspire and encourage employees to meet their quality goals, to make continual improvements, and to promote new ideas?	
95	6.3*	Production Supervisor/ Team Leader	What additional equipment and tools are needed in your area?	
96	6.4.1	Production Supervisor/ Team Leader	How do you address product safety and the ways to reduce possible risks to personnel?	
97	6.4.2	Production Supervisor/ Team Leader	How do you keep the production area orderly, clean, and in good state of repair?	

Auditor Name _____   Date of Audit _____

Section 3.6 Production Process

	ISO/TS 16949:2002 Clause * indicates ISO 9001:2000	Function/Level	Question	Result
98	7.2.2*	Production Supervisor/ Team Leader	When a customer order is amended, how is this communicated with production?	
99	7.2.2*	Production Supervisor/ Team Leader	When customer requirements are changed, how is this communicated with production?	
100	7.5.1*	Production Supervisor/ Team Leader	How are production processes controlled?	

Auditor Name _____   Date of Audit _____

Section 3.6 Production Process

	ISO/TS 16949:2002 Clause * indicates ISO 9001:2000	Function/Level	Question	Result
101	7.5.1.a*	Production Supervisor/ Team Leader	Where do you find information about product characteristics?	
102	7.5.1.b*/ 7.5.1.2	Production Supervisor/ Team Leader	Show me your work instructions and where they are located.	
103	7.5.1.c*	Production Supervisor/ Team Leader	How do you ensure your area is using suitable equipment?	
104	7.5.1.1	Production Supervisor/ Team Leader	Show me your control plans.	

Auditor Name _____   Date of Audit _____

Section 3.6 Production Process

	ISO/TS 16949:2002 Clause * indicates ISO 9001:2000	Function/Level	Question	Result
105	7.5.1.3	Production Supervisor/ Team Leader	When and how are job setups verified?	
106	7.5.2*/ 7.5.2.1	Production Supervisor/ Team Leader	How do you validate and revalidate processes? *(The auditor should sample each process to ensure it is properly controlled:* 1. *The process must be reviewed and approved based on defined criteria.* 2. *The equipment must be approved.* 3. *The persons working on it must be qualified.* 4. *There must be work instructions of some type.* 5. *The process must be revalidated.* 6. *There should be records for each of the above.)*	

Auditor Name _____   Date of Audit _____

Section 3.6 Production Process

	ISO/TS 16949:2002 Clause * indicates ISO 9001:2000	Function/Level	Question	Result
107	7.5.3*/ Note	Production Supervisor/ Team Leader	How is product identified in your areas?	
108	7.5.4*/ Note	Production Supervisor/ Team Leader	Do you have any customer property? If yes, how is it controlled? If it is lost, damaged, or otherwise unsuitable for use, how is this reported to the customer? What is the record?	
109	7.5.5*	Production Supervisor/ Team Leader	How is product preserved during production?	
110	7.5.5*	Production Supervisor/ Team Leader	What controls do you have in place for proper handling, packaging, storage, and protection of product?	

Auditor Name _____   Date of Audit _____

Section 3.6 Production Process

	ISO/TS 16949:2002 Clause * indicates ISO 9001:2000	Function/Level	Question	Result
111	8.1.2	Production Supervisor/ Team Leader	Thinking about statistical process control, what does stability mean? Process capability? Overadjustment?	
112	8.2.1*/ Note	Production Supervisor/ Team Leader	Who are your internal customers and how do you measure their satisfaction?	
113	8.2.3.1	Production Supervisor/ Team Leader	What should the production associate do if a process is found unstable or not capable?	
114	8.3	Production Supervisor/ Team Leader	What is your process to control nonconforming product?	

Auditor Name _____   Date of Audit _____

Section 3.6 Production Process

	ISO/TS 16949:2002 Clause * indicates ISO 9001:2000	Function/Level	Question	Result
115	8.3	Production Supervisor/ Team Leader	Do you rework product? If so, how is it reverified? *(Sample to ensure this is done.)*	
116	8.5.2/ 8.5.3*	Production Supervisor/ Team Leader	Have you participated in any corrective and/or preventive actions? If yes, please give an example.	
117	5.1.a*	Production Associate	How does the company communicate the importance of meeting customer, regulatory, and legal requirements to you?	
118	5.3.d*	Production Associate	What does the quality policy mean to you?	

Auditor Name _____   Date of Audit _____

Section 3.6 Production Process

	ISO/TS 16949:2002 Clause * indicates ISO 9001:2000	Function/Level	Question	Result
119	5.4.1*	Production Associate	What are the quality objectives for your area?	
120	5.5.1*	Production Associate	What are your primary responsibilities? *(If not documented, ensure this matches answer from manager.)*	
121	5.5.1*	Production Associate	Whom do you work for? *(Ensure this matches answer from manager.)*	
122	5.5.1.1	Production Associate	Who has the ability to stop production to fix quality issues?	

Auditor Name _____   Date of Audit _____

Section 3.6 Production Process

	ISO/TS 16949:2002 Clause * indicates ISO 9001:2000	Function/Level	Question	Result
123	5.5.1.1	Production Associate	Who is responsible to ensure product quality on your shift? *(Note: Auditor to confirm that the person responsible to ensure product quality also has the authority to stop production to correct quality problems.)*	
124	5.5.3*	Production Associate	How effective is your quality management system? How is this communicated to you?	
125	6.2.2.d*	Production Associate	How do you contribute to the achievement of the quality objectives?	
126	6.2.2.d*	Production Associate	What is the importance of your job?	

Auditor Name _____   Date of Audit _____

Section 3.6 Production Process

	ISO/TS 16949:2002 Clause * indicates ISO 9001:2000	Function/Level	Question	Result
127	6.2.2.2	Production Associate	How were you qualified to perform this job?	
128	6.2.2.2/ 6.2.2.3	Production Associate	What classroom training and on-the-job training have you received?	
129	6.2.2.3	Production Associate	For the part or material you are producing, if a nonconformity is shipped, what kind of problems would that create for the customer?	
130	6.2.2.4	Production Associate	What process does the company have to inspire employees to meet their quality objectives, to make continual improvements, and to encourage new ideas?	

Auditor Name _____   Date of Audit _____

Section 3.6 Production Process

	ISO/TS 16949:2002 Clause * indicates ISO 9001:2000	Function/Level	Question	Result
131	6.3*	Production Associate	What additional equipment and tools do you need to do your job?	
132	6.4.1	Production Associate	How does the company address product safety and the ways to reduce possible risks to personnel?	
133	6.4.2	Production Associate	How do you keep the production area orderly, clean, and in good state of repair?	
134	8.5.2/ 8.5.3*	Production Associate	Have you participated in any corrective and/or preventive actions? If yes, please give an example.	

Auditor Name _____   Date of Audit _____

Section 3.6 Production Process

	ISO/TS 16949:2002 Clause * indicates ISO 9001:2000	Function/Level	Question	Result
135	7.5.1	Production Associate	What process parameters do you monitor? What records do you keep? *(Auditor should verify that actual process settings match requirements. Use additional sheets if necessary.)* Parameter Actual Setting Required Setting 1. 2. 3. 4. 5. 6. 7. 8.	
136	7.5.1.b*/ 7.5.1.2	Production Associate	May I see your work instructions?	

Auditor Name _____   Date of Audit _____

Section 3.6 Production Process

	ISO/TS 16949:2002 Clause * indicates ISO 9001:2000	Function/Level	Question	Result
137	7.5.1.1	Production Associate	*(Auditor should take the control plan and ensure each step is completely followed. Ensure work instructions do not conflict with control plan.)*	
138	7.5.1.3	Production Associate	Show me the work instructions for the job setup. Show me how this setup was verified.	
139	7.5.3*/ Note	Production Associate	How is the product you are running identified?	

Auditor Name _____   Date of Audit _____

Section 3.6 Production Process

	ISO/TS 16949:2002 Clause * indicates ISO 9001:2000	Function/Level	Question	Result
140	7.5.5*	Production Associate	Does the product that you're running (name of product: _____) have any requirements for: Handling? Packaging? Storage? Protection?	
141	7.6	Production Associate	For the product that you're running (name of product: _____), what monitoring and measuring devices do you use? Be certain to include both product and process measuring devices, as applicable. *(Record these instruments on Appendix D, Calibration.)*	

Auditor Name _____   Date of Audit _____

Section 3.6 Production Process

	ISO/TS 16949:2002 Clause * indicates ISO 9001:2000	Function/Level	Question	Result
142	7.6	Production Associate	How do you know this measuring device is acceptable to use?	
143	8.1.2	Production Associate	*(If associate is using SPC, ask:)* What does process capability mean?	
144	8.1.2	Production Associate	*(If associate is using SPC, ask:)* How do you know if this process is stable?	
145	8.1.2	Production Associate	What is meant by the term "overadjustment"?	

Auditor Name _____ Date of Audit _____

Section 3.6 Production Process

	ISO/TS 16949:2002 Clause * indicates ISO 9001:2000	Function/Level	Question	Result
146	8.2.	Production Associate	For the product that you're running (name of product: _____), do you monitor and/or measure the product? *(Verify answer is correct and that measurements were taken at appropriate frequency and were within spec. Ensure answer is consistent with control plan requirements.)*	
147	8.2.3.1	Production Associate	What do you do if your process is no longer capable? If it is unstable?	

Auditor Name _____ Date of Audit _____

Section 3.6 Production Process

	ISO/TS 16949:2002 Clause * indicates ISO 9001:2000	Function/Level	Question	Result
148	8.2.3.1	Production Associate	*(Auditor should sample manufacturing processes to ensure they have been stable and capable since the last audit. If a process is found that has been unstable or not capable, ensure the reaction plan was followed, including containment of product and 100% inspection. Verify that a corrective action was completed.)*	
149	8.3	Production Associate	What do you do if you find product out of specification? What happens to the product?	
150	8.3.1	Production Associate	What do you do with product if it is unidentified? Suspect?	

Auditor Name _____   Date of Audit _____

Section 3.6 Production Process

	ISO/TS 16949:2002 Clause * indicates ISO 9001:2000	Function/Level	Question	Result
151	8.3.2	Production Associate	Do you rework product? If so, show me your rework instructions.	
152	8.5.2/ 8.5.3*	Production Associate	Have you participated in any corrective and/or preventive actions? If yes, please give an example.	
153	6.1*	Determine by observation in Production	*(Do resources appear to be adequate to maintain the quality system and continually improve its effectiveness? ... to enhance customer satisfaction by meeting customer requirements?)*	
154	6.3*	Determine by observation in Production	*(Infrastructure: Does the area include the appropriate workspace, equipment, and supporting services such as transport or communication?)*	

Auditor Name _____   Date of Audit _____

Section 3.6 Production Process

	ISO/TS 16949:2002 Clause * indicates ISO 9001:2000	Function/Level	Question	Result
155	6.4*	Determine by observation in Production	(Is the area's work environment appropriate to meet product requirements?)	
156	4.2.3*	All—Production	(Randomly sample at least 12 documents in the area to ensure they meet requirements of document control procedures. Be sure to include the quality manual, procedures, work instructions, forms, product specifications/drawings, and external documents. Use the Appendix B documents form.)	
157	4.2.4*/ 4.2.4.1	All—Production	(Randomly sample at least eight records in the area to ensure they meet requirements of record control procedures. Use the Appendix C records form.)	
158	4.2.3*	All—Lab	(Randomly sample at least 12 documents in the area to ensure they meet requirements of document control procedures. Be sure to include the quality manual, procedures, work instructions, forms, product specifications/drawings, and external documents. Use the Appendix B documents form.)	

Auditor Name _____   Date of Audit _____

Section 3.6 Production Process

	ISO/TS 16949:2002 Clause * indicates ISO 9001:2000	Function/Level	Question	Result
159	4.2.4*/ 4.2.4.1	All—Lab	(Randomly sample at least eight records in the area to ensure they meet requirements of record control procedures. Use the Appendix C records form.)	
160	6.1*	Determine by observation in Lab	(Do resources appear to be adequate to maintain the quality system and continually improve its effectiveness? ... to enhance customer satisfaction by meeting customer requirements?)	
161	6.3*	Determine by observation in Lab	(Infrastructure: Does the area include the appropriate workspace, equipment, and supporting services such as transport or communication?)	
162	6.4*	Determine by observation in Lab	(Is the area's work environment appropriate to meet product requirements?)	
163	4.1.a*	Quality/Lab Manager	What process(es) do you manage?	

Auditor Name _____   Date of Audit _____

Section 3.6 Production Process

	ISO/TS 16949:2002 Clause * indicates ISO 9001:2000	Function/Level	Question	Result
164	4.1.b*	Quality/Lab Manager	How does your process(es) link with the other processes in the company?	
165	4.2.3.1	Quality/Lab Manager	How do you make sure customer specifications and changes are examined, handed out, and applied in a timely manner? How do you make certain the review occurs within two weeks?	
166	5.1.a*	Quality/Lab Manager	How do you communicate the importance of meeting customer, regulatory, and legal requirements to your area?	
167	5.3.d*	Quality/Lab Manager	How do you ensure the quality policy is communicated and understood?	

Auditor Name _____   Date of Audit _____

Section 3.6 Production Process

	ISO/TS 16949:2002 Clause * indicates ISO 9001:2000	Function/Level	Question	Result
168	5.4.1*	Quality/Lab Manager	What are the quality objectives for your area?	
169	5.4.2*	Quality/Lab Manager	How do you ensure the integrity of the quality management system is maintained when changes are planned and implemented?	
170	5.5.1*	Quality/Lab Manager	Where are responsibilities and authorities defined, and how are they communicated?	
171	5.5.1.1	Quality/Lab Manager	Who can stop production to correct quality issues?	

Auditor Name _____   Date of Audit _____

Section 3.6 Production Process

	ISO/TS 16949:2002 Clause * Indicates ISO 9001:2000	Function/Level	Question	Result
172	5.5.1.1	Quality/Lab Manager	Who is responsible to ensure product quality on each shift? *(Note: Auditor to confirm that the person responsible to ensure product quality also has the authority to stop production to correct quality problems.)*	
173	5.5.2.1	Quality/Lab Manager	Who has been given the responsibility and authority to make sure customer requirements are met? *(If quality manager, ask how this was done. If someone else, make certain this question is asked to the appropriate person.)*	
174	5.5.3*	Quality/Lab Manager	What communication processes are established within your area?	
175	5.5.3*	Quality/Lab Manager	How do you communicate the effectiveness of your quality management system to your area?	

Auditor Name _____   Date of Audit _____

Section 3.6 Production Process

	ISO/TS 16949:2002 Clause * Indicates ISO 9001:2000	Function/Level	Question	Result
176	6.2.1*/ 7.6.3.1	Quality/Lab Manager	How do you determine competence of your employees?	
177	6.2.2.d*	Quality/Lab Manager	How do you ensure that personnel in your area are aware of the importance of their job and how they contribute to achievement of quality objectives?	
178	6.2.2.2	Quality/Lab Manager	How do you qualify employees?	
179	6.2.2.3	Quality/Lab Manager	What on-the-job training is provided for your new employees? ... for employees that have changed jobs? ... for contract or agency personnel?	

Auditor Name _____   Date of Audit _____

Section 3.6 Production Process

	ISO/TS 16949:2002 Clause * indicates ISO 9001:2000	Function/Level	Question	Result
180	6.2.2.3	Quality/Lab Manager	How do you inform your employees of the consequences to the customer of nonconformity to quality requirements?	
181	6.2.2.4	Quality/Lab Manager	What is your process to inspire employees to meet their quality goals, to make continual improvements, and to encourage new ideas?	
182	7.1	Quality/Lab Manager	What is your role in quality planning?	
183	7.1	Quality/Lab Manager	Do any of your customers require you to follow the APQP AIAG Reference Manual[13] or another product approval process? If so, please describe how this is done.	

Auditor Name _____   Date of Audit _____

Section 3.6 Production Process

	ISO/TS 16949:2002 Clause * indicates ISO 9001:2000	Function/Level	Question	Result
184	7.1.c*	Quality/Lab Manager	During planning, how do you determine required verification, validation, monitoring, inspection, and test activities specific to the product?	
185	7.1.c*/ 7.1.2	Quality/Lab Manager	During planning, how do you determine the criteria for product acceptance? What are your requirements for customer approvals?	
186	7.1	Quality/Lab Manager	What is the output of the planning process?	
187	7.1.1	Quality/Lab Manager	How do you make sure customer requirements are included in the planning process as a component of the quality plan?	
188	7.1.2	Quality/Lab Manager	What is your acceptance level for attribute sampling data?	

Auditor Name _____   Date of Audit _____

Section 3.6 Production Process

	ISO/TS 16949:2002 Clause * indicates ISO 9001:2000	Function/Level	Question	Result
189	7.1.4	Quality/Lab Manager	How are product changes controlled?	
190	7.2.2.2	Quality/Lab Manager	What is your role in determining manufacturing viability of planned products?	
191	7.3.1.1/ Note	Quality/Lab Manager	What is your role in the design process?	
192	7.3.6.2	Quality/Lab Manager	What is your role in the prototype program?	
193	7.3.6.3	Quality/Lab Manager	If any of the prototype program is outsourced, who is responsible for the outsourced service, including technical leadership?	

Auditor Name _____   Date of Audit _____

Section 3.6 Production Process

	ISO/TS 16949:2002 Clause * indicates ISO 9001:2000	Function/Level	Question	Result
194	7.3.6.3	Quality/Lab Manager	*(Auditor to sample products to ensure the customer's product approval procedure was used.)*	
195	7.4.3.1	Quality/Lab Manager	What is your process for incoming product quality?	
196	7.6	Quality/Lab Manager	Who is responsible for calibrations/verifications in the lab? In the plant?	
197	7.6	Quality/Lab Manager	What software is used in the monitoring and measurement of specified requirements? *(Auditor to verify that software was confirmed prior to initial use and reconfirmed as necessary.)*	

Auditor Name _____   Date of Audit _____

Section 3.6 Production Process

	ISO/TS 16949:2002 Clause * Indicates ISO 9001:2000	Function/Level	Question	Result
198	7.6.1	Quality/Lab Manager	*(Auditor to verify that a variation study has been conducted for each type of measuring and test equipment referenced in the control plans.)*	
199	7.6.3.1	Quality/Lab Manager	*(Is the lab accredited to ISO/IEC 17025? If no, move to next question. If yes, view certificate to ensure the accreditation covers inspection and tests performed. If complete, skip the next two questions.)*	
200	7.6.3.1	Quality/Lab Manager	Show me your lab scope.	

Auditor Name _____ Date of Audit _____

Section 3.6 Production Process

	ISO/TS 16949:2002 Clause * Indicates ISO 9001:2000	Function/Level	Question	Result
201	7.6.3.1	Quality/Lab Manager	*(Auditor to verify that the lab has specified and implemented:* *- Sufficient lab procedures* *- Capable lab employees* *- Testing of product* *- Ability to correctly perform these services, traceable to appropriate process standard* *- Review of records)*	

Auditor Name _____ Date of Audit _____

Section 3.6 Production Process

	ISO/TS 16949:2002 Clause * indicates ISO 9001:2000	Function/Level	Question	Result
202	7.6.3.2	Quality/Lab Manager	Do you use any external labs for inspection, test, or calibration services pertaining to the design process? *(If yes, verify that these labs are either acceptable to the customer, are the OEM, or are accredited.)*	
203	8.1.1	Quality/Lab Manager	How do you make sure statistical tools are determined during advanced quality planning and included in the control plan?	
204	8.1.2	Quality/Lab Manager	How do you make sure basic statistical techniques are understood and utilized?	
205	8.2.1*	Quality/Lab Manager	How do you obtain and use information pertaining to customer satisfaction?	

Auditor Name _____   Date of Audit _____

Section 3.6 Production Process

	ISO/TS 16949:2002 Clause * indicates ISO 9001:2000	Function/Level	Question	Result
206	8.2.1/ Note*	Quality/Lab Manager	Who are your internal customers and how do you measure their satisfaction?	
207	8.2.1.1	Quality/Lab Manager	Show me performance indicators for delivered part quality performance, customer disruptions including field returns, delivery schedule performance, premium freight, and customer notifications related to quality or delivery issues as required by the standard.	
208	8.2.2.3	Quality/Lab Manager	Explain your process to audit products at suitable stages of production and delivery to confirm conformity to all specific requirements at a definite interval. These requirements may include dimensions, functionality, packaging, and labeling.	
209	8.2.3*	Quality/Lab Manager	What do you do when planned results are not achieved? *(Examples of correction and corrective action should be provided.)*	

Auditor Name _____   Date of Audit _____

Section 3.6 Production Process

	ISO/TS 16949:2002 Clause * indicates ISO 9001:2000	Function/Level	Question	Result
210	8.2.3*/ 4.1.e*	Quality/Lab Manager	How do you monitor and measure your process? Please provide examples.	
211	8.2.3.1	Quality/Lab Manager	How do you ensure that manufacturing process capability specified on the product approval process is sustained? *(Auditor should verify that the capability is being monitored.)*	
212	8.2.4*/ Note	Quality/Lab Manager	Show me the record that indicates who authorized release of product.	
213	8.2.4*/ Note	Quality/Lab Manager	How do you make certain product release does not occur until all inspections and tests have been satisfactorily completed?	
214	8.2.4.1	Quality/Lab Manager	What is your process for layout inspection and functional testing?	

Auditor Name _____   Date of Audit _____

Section 3.6 Production Process

	ISO/TS 16949:2002 Clause * indicates ISO 9001:2000	Function/Level	Question	Result
215	8.2.4.2	Quality/Lab Manager	Do you have any parts designated as "appearance items" by your customer? If so, what are they?	
216	8.3	Quality/Lab Manager	What is your process to control nonconforming product?	
217	8.3	Quality/Lab Manager	How is nonconforming product identified and tracked?	
218	8.3	Quality/Lab Manager	What is your record of nonconformity and subsequent actions taken, including concessions?	

Auditor Name _____   Date of Audit _____

Section 3.6 Production Process

	ISO/TS 16949:2002 Clause * indicates ISO 9001:2000	Function/Level	Question	Result
219	8.3	Quality/Lab Manager	What do you do if nonconforming product is found after delivery or use has started?	
220	8.3.1	Quality/Lab Manager	How do you handle unidentified or suspect product?	
221	8.3.1	Quality/Lab Manager	Show me reinspection requirements in rework instructions.	
222	8.3.1	Quality/Lab Manager	If nonconforming product is shipped, how is the customer punctually notified?	
223	8.3.4	Quality/Lab Manager	What is your process for customer waivers?	

Auditor Name _____   Date of Audit _____

Section 3.6 Production Process

	ISO/TS 16949:2002 Clause * indicates ISO 9001:2000	Function/Level	Question	Result
224	8.3.4	Quality/Lab Manager	Show me your records of customer waivers. *(Auditor to ensure expiration date or quantity authorized is recorded.)*	
225	8.3.4	Quality/Lab Manager	How do you identify shipping containers for material shipped on a customer waiver?	
226	8.4.b*	Quality/Lab Manager	Show me where data is analyzed to show conformity to product requirements.	
227	8.4.b*	Quality/Lab Manager	Show me where data is analyzed for characteristics and trends of products including opportunities for preventive action.	

Auditor Name _____   Date of Audit _____

Section 3.6 Production Process

	ISO/TS 16949:2002 Clause * indicates ISO 9001:2000	Function/Level	Question	Result
228	8.5.1*	Quality/Lab Manager	Please provide examples of continual improvement in your area(s) of responsibility.	
229	8.5.1.2	Quality/Lab Manager	Show me evidence that the company constantly concentrates on control and reduction of variation in product characteristics.	
230	8.5.2	Quality/Lab Manager	What internal corrective actions have been issued/assigned to or closed in your area(s) since the last internal audit? *(Auditor should randomly sample at least six corrective actions.)*	

Auditor Name _____   Date of Audit _____

Section 3.6 Production Process

	ISO/TS 16949:2002 Clause * indicates ISO 9001:2000	Function/Level	Question	Result
231	8.5.2	Quality/Lab Manager	What customer complaints have been issued/assigned to or closed in your area(s) since the last internal audit? *(Auditor should randomly sample at least six customer complaints.)*	
232	8.5.2.4	Quality/Lab Manager	How do you make certain customer returns are analyzed promptly and records maintained?	
233	8.5.2.4	Quality/Lab Manager	Show me the records of customer return analysis, including corrective action to prevent recurrence.	

Auditor Name _____   Date of Audit _____

Section 3.6 Production Process

	ISO/TS 16949:2002 Clause * indicates ISO 9001:2000	Function/Level	Question	Result
234	8.5.3*	Quality/Lab Manager	What preventive actions have been issued or closed in your area(s) since the last internal audit? *(Because this number is usually small, the auditor should review all preventive actions.)*	
235	5.1.a*	Lab Supervisor/ Team Leader	How do you communicate the importance of meeting customer, regulatory, and legal requirements to your area?	
236	5.3.d*	Lab Supervisor/ Team Leader	What does the quality policy mean to you? How do you ensure the quality policy is communicated and understood?	

Auditor Name _____   Date of Audit _____

Section 3.6 Production Process

	ISO/TS 16949:2002 Clause * indicates ISO 9001:2000	Function/Level	Question	Result
237	5.4.1*	Lab Supervisor/ Team Leader	What are the quality objectives for your area?	
238	5.5.1.1	Lab Supervisor/ Team Leader	Who can stop production to fix quality issues?	
239	5.5.1.1	Lab Supervisor/ Team Leader	Who is responsible to ensure product quality on your shift? *(Note: Auditor to confirm that the person responsible to ensure product quality also has the authority to stop production to correct quality problems.)*	
240	5.5.3*	Lab Supervisor/ Team Leader	What communication processes do you use?	

Auditor Name _____   Date of Audit _____

Section 3.6 Production Process

	ISO/TS 16949:2002 Clause * indicates ISO 9001:2000	Function/Level	Question	Result
241	5.5.3*	Lab Supervisor/ Team Leader	How effective is your quality management system? How do you know?	
242	6.2.1*/ 7.6.3.1	Lab Supervisor/ Team Leader	How do you determine competence of your employees?	
243	6.2.2.d*	Lab Supervisor/ Team Leader	How do you ensure that personnel in your area are aware of the importance of their jobs and how they contribute to achievement of quality objectives?	
244	6.2.2.d*	Lab Supervisor/ Team Leader	How do you contribute to the achievement of the quality objectives?	

Auditor Name _____   Date of Audit _____

Section 3.6 Production Process

	ISO/TS 16949:2002 Clause * indicates ISO 9001:2000	Function/Level	Question	Result
245	6.2.2.d*	Lab Supervisor/ Team Leader	What is the importance of your job?	
246	6.2.2.2	Lab Supervisor/ Team Leader	How do you qualify employees?	
247	6.2.2.3	Lab Supervisor/ Team Leader	What on-the-job training is provided for your new employees? … for employees that have changed jobs? … for contract or agency personnel?	
248	6.2.2.3	Lab Supervisor/ Team Leader	How do you inform your employees about the penalty of nonconformities to your customers? Please provide specific examples.	

Auditor Name _____   Date of Audit _____

Section 3.6 Production Process

	ISO/TS 16949:2002 Clause * indicates ISO 9001:2000	Function/Level	Question	Result
249	6.2.2.4	Lab Supervisor/ Team Leader	What is your process to inspire employees to meet their quality goals, to make continual improvements, and to encourage new ideas?	
250	6.3*	Lab Supervisor/ Team Leader	What additional tools and equipment are needed in your area?	
251	7.1.4	Lab Supervisor/ Team Leader	How are changes controlled in the lab?	
252	7.5.3*/ Note	Lab Supervisor/ Team Leader	How are samples and/or parts identified while in the lab?	
253	7.6/ 7.6.2/ 7.6.3.2	Lab Supervisor/ Team Leader	Do you use any outside calibration companies? *(If yes, add to list of suppliers on Appendix A, Purchasing.)*	

Auditor Name _____   Date of Audit _____

Section 3.6 Production Process

	ISO/TS 16949:2002 Clause * indicates ISO 9001:2000	Function/Level	Question	Result
254	7.6	Lab Supervisor/ Team Leader	*(For samples collected on Appendix D, Calibration, that the lab is responsible for, the auditor should ensure all calibration requirements are met for each instrument. Be sure each instrument was checked against standards traceable to international or national measurement standards. Or if there is no such standard, the basis for calibration/verification must be recorded. Add internal standards used to Appendix D, Calibration. Be sure to sample both instruments calibrated and verified internally and those calibrated by an outside supplier. If any other outside calibration suppliers are found but not mentioned previously during the audit, be sure to add them to Appendix A, Purchasing.)*	
255	7.6.2	Lab Supervisor/ Team Leader	Make sure records noted in previous question meet the requirements of 7.6.2.	

Auditor Name _____   Date of Audit _____

Section 3.6 Production Process

	ISO/TS 16949:2002 Clause * indicates ISO 9001:2000	Function/Level	Question	Result
256	7.6*	Lab Supervisor/ Team Leader	How do you ensure monitoring and measurement devices are protected during handling, maintenance, and storage?	
257	7.6	Lab Supervisor/ Team Leader	What do you do if an instrument does not conform to requirements? *(The auditor must verify that the validity of previous measuring results is assessed and recorded. There should be evidence that the proper action was taken on the instrument and any product affected.)*	
258	8.1.2	Lab Supervisor/ Team Leader	What is process capability? Control or stability?	

Auditor Name _____   Date of Audit _____

Section 3.6 Production Process

	ISO/TS 16949:2002 Clause * indicates ISO 9001:2000	Function/Level	Question	Result
259	8.2.2.3	Lab Supervisor/ Team Leader	What product audits are conducted? *(Auditor to sample and verify records.)*	
260	8.2.4*/ Note	Lab Supervisor/ Team Leader	What records show evidence of conformity with the acceptance criteria and the person authorizing release? *(Auditor to sample and verify.)*	
261	8.2.4.1	Lab Supervisor/ Team Leader	*(Auditor should sample at least six parts and verify that layout inspection and functional verification were performed as specified in the control plan.)*	

Auditor Name _____   Date of Audit _____

Section 3.6 Production Process

	ISO/TS 16949:2002 Clause * Indicates ISO 9001:2000	Function/Level	Question	Result
262	8.2.4.2	Lab Supervisor/ Team Leader	*(If applicable, auditor should sample at least six parts designated as appearance items by the customer and verify that requirements of 8.2.4.2 are met.)*	
263	8.3	Lab Supervisor/ Team Leader	What is your process to control nonconforming product?	

Auditor Name _____   Date of Audit _____

Section 3.6 Production Process

	ISO/TS 16949:2002 Clause * Indicates ISO 9001:2000	Function/Level	Question	Result
264	8.3	Lab Supervisor/ Team Leader	*(Auditor should randomly select at least six products identified as nonconforming product since the last internal audit. Ensure procedures for nonconforming product were followed for each. Ensure appropriate concessions were obtained if applicable.)* 1. 2. 3. 4. 5. 6.	
265	8.3	Lab Supervisor/ Team Leader	What nonconforming product is currently in-house? *(Auditor should randomly choose nonconforming product and ensure it is properly identified.)*	

Auditor Name _____   Date of Audit _____

Section 3.6 Production Process

	ISO/TS 16949:2002 Clause * Indicates ISO 9001:2000	Function/Level	Question	Result
266	8.5.2	Lab Supervisor/ Team Leader	What internal corrective actions have you been involved with since the last internal audit? *(Auditor should randomly sample at least six corrective actions.)*	
267	8.5.2	Lab Supervisor/ Team Leader	What customer complaints have you participated in since the last internal audit? *(Auditor should randomly sample at least six customer complaints.)*	
268	5.1.a*	Lab Tech	How does the company communicate the importance of meeting customer, regulatory, and legal requirements to you?	

Auditor Name _____ Date of Audit _____

Section 3.6 Production Process

	ISO/TS 16949:2002 Clause * Indicates ISO 9001:2000	Function/Level	Question	Result
269	5.3.d*	Lab Tech	What does the quality policy mean to you?	
270	5.4.1*	Lab Tech	What are the quality objectives for your area?	
271	5.5.1*	Lab Tech	What are your primary responsibilities? *(If not documented, ensure this matches answer from manager.)*	
272	5.5.1*	Lab Tech	Whom do you work for? *(Ensure this matches answer from manager.)*	

Auditor Name _____ Date of Audit _____

Section 3.6 Production Process

	ISO/TS 16949:2002 Clause * indicates ISO 9001:2000	Function/Level	Question	Result
273	5.5.1.1	Lab Tech	Who can stop production to fix quality issues?	
274	5.5.1.1	Lab Tech	*Who is responsible to ensure product quality on your shift? (Note: auditor to confirm that the person responsible to ensure product quality also has the authority to stop production to correct quality problems.)*	
275	5.5.3*	Lab Tech	How effective is your quality management system? How is this communicated to you?	
276	6.2.2.d*	Lab Tech	How do you contribute to the achievement of the quality objectives?	

Auditor Name _____   Date of Audit _____

Section 3.6 Production Process

	ISO/TS 16949:2002 Clause * indicates ISO 9001:2000	Function/Level	Question	Result
277	6.2.2.d*	Lab Tech	What is the importance of your job?	
278	6.2.2.2/ 7.6.3.1	Lab Tech	How were you qualified to perform this job?	
279	6.2.2.2/ 6.2.2.3	Lab Tech	What classroom training and on-the-job training have you received?	
280	6.2.2.3	Lab Tech	For the part or material you are testing/inspecting, if a nonconformity is shipped, what kind of problems would that create for the customer?	

Auditor Name _____   Date of Audit _____

Section 3.6 Production Process

	ISO/TS 16949:2002 Clause * indicates ISO 9001:2000	Function/Level	Question	Result
281	6.2.2.4	Lab Tech	What process does the company have to inspire employees to meet their quality goals, to make continual improvements, and to encourage new ideas?	
282	6.3*	Lab Tech	What additional equipment and tools do you need to do your job?	
283	7.5.3*/ Note	Lab Tech	What sample or part are you working on now? How is it identified while in the lab? Name of product: _____	
284	7.6	Lab Tech	What monitoring and measuring devices are used for this product? *(Record the instruments on Appendix D, Calibration.)*	

Auditor Name _____   Date of Audit _____

Section 3.6 Production Process

	ISO/TS 16949:2002 Clause * indicates ISO 9001:2000	Function/Level	Question	Result
285	7.6	Lab Tech	How do you know these measuring devices are acceptable to use?	
286	7.6.3.1	Lab Tech	Show me your lab procedures for the tests required on this product.	
287	7.6.3.2	Lab Tech	If this product is sent to an outside lab for testing, what facility is used? *(Auditor to verify requirements for external lab met.)*	
288	8.1.2	Lab Tech	What is meant by process capability? Control or stability? Overadjustment?	

Auditor Name _____   Date of Audit _____

Section 3.6 Production Process

	ISO/TS 16949:2002 Clause * indicates ISO 9001:2000	Function/Level	Question	Result
289	8.2.3.1	Lab Tech	What do you do if you find a characteristic that is either not capable or unstable?	
290	8.2.4*/ Note	Lab Tech	What tests and/or measurements are required for this product? How do you know this?	

Auditor Name _____   Date of Audit _____

Section 3.6 Production Process

	ISO/TS 16949:2002 Clause * indicates ISO 9001:2000	Function/Level	Question	Result
291	8.2.4	Lab Tech	(Ask the lab tech to show you the records from previous test results. Randomly select records from at least six runs since the last internal audit [attach additional sheets if necessary]. Ensure all required tests were performed and all results were within specification.) 1. 2. 3. 4. 5. 6.	
292	8.3	Lab Tech	What do you do when product is found outside specification?	

Auditor Name _____   Date of Audit _____

Section 3.6 Production Process

	ISO/TS 16949:2002 Clause * Indicates ISO 9001:2000	Function/Level	Question	Result
293	8.5.2/ 8.5.3*	Lab Tech	Have you participated in any corrective and/or preventive actions? If yes, please give an example.	
294	4.2.4*/ 4.2.4.1	All— Maintenance	*(Randomly sample at least six records in the area to ensure they meet requirements of record control procedures. Use the Appendix C records form.)*	
295	4.2.3*	All— Maintenance	*(Randomly sample at least six documents in the area to ensure they meet requirements of document control procedures. Be sure to include the quality manual, procedures, work instructions, forms, and external documents. Use the Appendix B documents form.)*	
296	6.1*	Determine by observation in Maintenance	*(Do resources appear to be adequate to maintain the quality system and continually improve its effectiveness? ... to enhance customer satisfaction by meeting customer requirements?)*	

Auditor Name _____   Date of Audit _____

Section 3.6 Production Process

	ISO/TS 16949:2002 Clause * Indicates ISO 9001:2000	Function/Level	Question	Result
297	6.3*	Determine by observation in Maintenance	*(Infrastructure: Does the area include the appropriate workspace, equipment, and supporting services such as transport or communication?)*	
298	6.4*	Determine by observation in Maintenance	*(Is the area's work environment appropriate to meet product requirements?)*	
299	4.1.a*	Maintenance Manager	What process(es) do you manage?	

Auditor Name _____   Date of Audit _____

Section 3.6 Production Process

	ISO/TS 16949:2002 Clause *indicates ISO 9001:2000	Function/Level	Question	Result
300	8.2.3*/ 4.1.e*	Maintenance Manager	How do you monitor and measure your process? Please provide examples.	
301	7.5.1.4	Maintenance Manager	Please provide evidence that you document, assess, and improve maintenance goals.	
302	4.1.b*	Maintenance Manager	How does your process(es) link with the other processes in the company?	

Auditor Name _____   Date of Audit _____

Section 3.6 Production Process

	ISO/TS 16949:2002 Clause *indicates ISO 9001:2000	Function/Level	Question	Result
303	5.1.a*	Maintenance Manager	How do you communicate the importance of meeting customer, regulatory, and legal requirements to your area?	
304	5.3.d*	Maintenance Manager	How do you ensure the quality policy is communicated and understood?	
305	5.4.1*	Maintenance Manager	What are the quality objectives for your area?	

Auditor Name _____   Date of Audit _____

Section 3.6 Production Process

	ISO/TS 16949:2002 Clause * Indicates ISO 9001:2000	Function/Level	Question	Result
306	5.5.1*	Maintenance Manager	Where are responsibilities and authorities defined, and how are they communicated?	
307	5.4.2*	Maintenance Manager	How do you ensure the integrity of the quality management system is maintained when changes are planned and implemented?	
308	5.5.3*	Maintenance Manager	What communication processes are established within your area?	

Auditor Name _____   Date of Audit _____

Section 3.6 Production Process

	ISO/TS 16949:2002 Clause * Indicates ISO 9001:2000	Function/Level	Question	Result
309	5.5.3*	Maintenance Manager	How do you communicate the effectiveness of your quality management system to your area?	
310	6.2.1*	Maintenance Manager	How do you determine competence of your employees?	
311	6.2.2.d*	Maintenance Manager	How do you ensure that personnel in your area are aware of the importance of their job and how they contribute to achievement of quality objectives?	
312	6.2.2.2/ 6.2.2.3	Maintenance Manager	What classroom and on-the-job training is required for maintenance personnel?	

Auditor Name _____   Date of Audit _____

Section 3.6 Production Process

	ISO/TS 16949:2002 Clause * indicates ISO 9001:2000	Function/Level	Question	Result
313	6.2.2.4	Maintenance Manager	What is your process to inspire employees to meet quality goals, to continually improve, and to encourage new ideas?	
314	7.1*	Maintenance Manager	How does maintenance participate in quality planning?	
315	7.3	Maintenance Manager	What is maintenance's role in manufacturing process design?	
316	7.5.1.4	Maintenance Manager	Show me where you've identified your key process equipment.	

Auditor Name _____   Date of Audit _____

Section 3.6 Production Process

	ISO/TS 16949:2002 Clause * indicates ISO 9001:2000	Function/Level	Question	Result
317	7.5.1.4	Maintenance Manager	Explain your process for maintenance/preventive maintenance, including planned maintenance activities.	
318	7.5.1.4	Maintenance Manager	*(Auditor should sample all production areas for required preventive and predictive maintenance. Include any new equipment that is in operation. Ensure PMs are being performed as required.)*	
319	7.5.1.4	Maintenance Manager	Show me your past due PMs and how you manage them.	

Auditor Name _____   Date of Audit _____

Section 3.6 Production Process

	ISO/TS 16949:2002 Clause * indicates ISO 9001:2000	Function/Level	Question	Result
320	7.5.1.4	Maintenance Manager	What is your system for packaging and preservation of equipment, tooling, and gauging?	
321	7.5.1.4	Maintenance Manager	What is your system to ensure availability of replacement parts for important manufacturing equipment?	
322	7.5.1.5	Maintenance Manager	Explain your system for production tooling management. *(Auditor should ensure all requirements of 7.5.1.5 are met.)*	
323	7.5.1.5	Maintenance Manager	How do you monitor any outsourced tooling management activities?	
324	7.5.4.1	Maintenance Manager	Show me how customer-owned tooling is permanently identified.	

Auditor Name _____   Date of Audit _____

Section 3.6 Production Process

	ISO/TS 16949:2002 Clause * indicates ISO 9001:2000	Function/Level	Question	Result
325	7.6	Maintenance Manager	What monitoring and measuring devices are controlled by maintenance?	
326	8.2.1*/ Note	Maintenance Manager	Who are your internal customers and how do you measure their satisfaction?	
327	8.2.3	Maintenance Manager	What do you do when planned results are not achieved? *(Examples of correction and corrective action should be provided.)*	
328	8.2.3.1	Maintenance Manager	Where are important process events such as tool change or machine repair recorded? *(Auditor to sample and verify.)*	

Auditor Name _____   Date of Audit _____

Section 3.6 Production Process

	ISO/TS 16949:2002 Clause * indicates ISO 9001:2000	Function/Level	Question	Result
329	8.5.1	Maintenance Manager	Please provide examples of continual improvement in your area(s) of responsibility.	
330	8.5.2	Maintenance Manager	What internal corrective actions have been issued/assigned to or closed in your area(s) since the last internal audit? *(Auditor should randomly sample at least six corrective actions.)*	
331	8.5.3*	Maintenance Manager	What preventive actions have been issued or closed in your area(s) since the last internal audit? *(Because this number is usually small, the auditor should review all preventive actions.)*	

Auditor Name _____   Date of Audit _____

Section 3.6 Production Process

	ISO/TS 16949:2002 Clause * indicates ISO 9001:2000	Function/Level	Question	Result
332	5.1.a*	Maintenance Supervisor/ Team Leader	How do you communicate the importance of meeting customer, regulatory, and legal requirements to your area?	
333	5.3.d*	Maintenance Supervisor/ Team Leader	What does the quality policy mean to you? How do you ensure the quality policy is communicated and understood?	
334	5.4.1*	Maintenance Supervisor/ Team Leader	What are the quality objectives for your area?	
335	5.5.3*	Maintenance Supervisor/ Team Leader	What communication processes do you use?	

Auditor Name _____   Date of Audit _____

Section 3.6 Production Process

	ISO/TS 16949:2002 Clause * indicates ISO 9001:2000	Function/Level	Question	Result
336	5.5.3*	Maintenance Supervisor/ Team Leader	How effective is your quality management system? How do you know?	
337	6.3*	Maintenance Supervisor/ Team Leader	What additional tools and equipment are needed in your area?	
338	6.2.2.d*	Maintenance Supervisor/ Team Leader	How do you contribute to the achievement of the quality objectives?	
339	6.2.2.d*	Maintenance Supervisor/ Team Leader	What is the importance of your job?	

Auditor Name _____   Date of Audit _____

Section 3.6 Production Process

	ISO/TS 16949:2002 Clause * indicates ISO 9001:2000	Function/Level	Question	Result
340	7.5.1.4	Maintenance Supervisor/ Team Leader	Explain your process for maintenance/preventive maintenance.	
341	7.5.1.4	Maintenance Supervisor/ Team Leader	What do you do when work orders are past due?	
342	7.5.1.5/ 7.5.4.1	Maintenance Supervisor/ Team Leader	*(Auditor should sample tooling to ensure requirements of 7.5.1.5 are met. Verify customer-owned tooling is permanently marked.)*	

Auditor Name _____   Date of Audit _____

Section 3.6 Production Process

	ISO/TS 16949:2002 Clause * indicates ISO 9001:2000	Function/Level	Question	Result
343	7.6	Maintenance Supervisor/ Team Leader	*(For samples collected on Appendix D, Calibration, that maintenance is responsible for, the auditor should ensure all calibration requirements are met for each instrument. Be sure each instrument was checked against standards traceable to international or national measurement standards. Or if there is no such standard, the basis for calibration/verification must be recorded. Add internal standards used to Appendix D, Calibration. Be sure to sample both instruments calibrated and verified internally and those calibrated by an external supplier. Add external providers of calibration services to Appendix A, Purchasing.)*	
344	7.6/ 7.5.1.4	Maintenance Supervisor/ Team Leader	How do you ensure monitoring and measurement devices are protected during handling, maintenance, and storage?	

Auditor Name _____   Date of Audit _____

Section 3.6 Production Process

	ISO/TS 16949:2002 Clause * indicates ISO 9001:2000	Function/Level	Question	Result
345	7.6	Maintenance Supervisor/ Team Leader	What do you do if an instrument does not conform to requirements? *(The auditor must verify that the validity of previous measuring results is assessed and recorded. There should be evidence that the proper action was taken on the instrument and any product affected.)*	
346	5.1.a*	Maintenance Tech	How does the company communicate the importance of meeting customer, regulatory, and legal requirements to you?	
347	5.3.d*	Maintenance Tech	What does the quality policy mean to you?	
348	5.4.1*/ 7.5.1.4	Maintenance Tech	What are the quality/maintenance objectives for your area?	

Auditor Name _____   Date of Audit _____

Section 3.6 Production Process

	ISO/TS 16949:2002 Clause * indicates ISO 9001:2000	Function/Level	Question	Result
349	5.5.1*	Maintenance Tech	What are your primary responsibilities? *(If not documented, ensure this matches answer from manager.)*	
350	5.5.1*	Maintenance Tech	Whom do you work for? *(Ensure this matches answer from manager.)*	
351	5.5.3*	Maintenance Tech	How effective is your quality management system? How is this communicated to you?	
352	6.3*	Maintenance Tech	What additional equipment and tools do you need to do your job?	
353	6.2.2.d*	Maintenance Tech	How do you contribute to the achievement of the quality objectives?	

Auditor Name _____   Date of Audit _____

Section 3.6 Production Process

	ISO/TS 16949:2002 Clause * indicates ISO 9001:2000	Function/Level	Question	Result
354	6.2.2.d*	Maintenance Tech	What is the importance of your job?	
355	6.2.2.2/ 6.2.2.3	Maintenance Tech	What training have you received?	
356	6.2.2.4	Maintenance Tech	What process does the company have to inspire employees to meet quality goals, to continually improve, and to encourage new ideas?	
357	7.5.1.4	Maintenance Tech	How do you know what is required to complete a particular PM?	

Auditor Name _____   Date of Audit _____

Section 3.6 Production Process

	ISO/TS 16949:2002 Clause * Indicates ISO 9001:2000	Function/Level	Question	Result
358	7.6	Maintenance Tech	What calibrations and/or verifications do you perform on monitoring and measuring devices? For each, what procedures are used?	
359	8.5.2/ 8.5.3*	Maintenance Tech	Have you participated in any corrective and/or preventive actions? If yes, please give an example.	
360	4.1.a*	Shipping Manager	What process(es) do you manage?	
361	4.1.b*	Shipping Manager	How does your process(es) link with the other processes in the company?	

Auditor Name _____   Date of Audit _____

Section 3.6 Production Process

	ISO/TS 16949:2002 Clause * Indicates ISO 9001:2000	Function/Level	Question	Result
362	5.1.a*	Shipping Manager	How do you communicate the importance of meeting customer, regulatory, and legal requirements to your area?	
363	5.3.d*	Shipping Manager	How do you ensure the quality policy is communicated and understood?	
364	5.4.1*	Shipping Manager	What are the quality objectives for your area?	
365	5.4.2*	Shipping Manager	How do you ensure the integrity of the quality management system is maintained when changes are planned and implemented?	

Auditor Name _____   Date of Audit _____

Section 3.6 Production Process

	ISO/TS 16949:2002 Clause * indicates ISO 9001:2000	Function/Level	Question	Result
366	5.5.1*	Shipping Manager	How are responsibilities and authorities defined, and how are they communicated?	
367	5.5.3*	Shipping Manager	What communication processes are established within your area?	
368	5.5.3*	Shipping Manager	How do you communicate the effectiveness of your quality management system to your area?	
369	6.2.1*	Shipping Manager	How do you determine competence of your employees?	

Auditor Name _____   Date of Audit _____

Section 3.6 Production Process

	ISO/TS 16949:2002 Clause * indicates ISO 9001:2000	Function/Level	Question	Result
370	6.2.2.d*	Shipping Manager	How do you ensure that personnel in your area are aware of the importance of their jobs and how they contribute to achievement of quality objectives?	
371	6.2.2.2	Shipping Manager	How do you qualify employees?	
372	6.2.2.3	Shipping Manager	What on-the-job training is provided for your new employees? … for employees that have changed jobs? … for contract or agency personnel?	
373	6.2.2.3	Shipping Manager	How do you inform your employees about the penalty of nonconformities, including shipping, to your customers? Please provide specific examples.	

Auditor Name _____   Date of Audit _____

Section 3.6 Production Process

	ISO/TS 16949:2002 Clause * indicates ISO 9001:2000	Function/Level	Question	Result
374	6.2.2.4	Shipping Manager	What is your process to inspire employees to meet their quality goals, to make continual improvements, and to encourage new ideas?	
375	7.2.2*	Shipping Manager	How do you make certain customer requirements such as packaging, labeling, and timing are met in shipping?	
376	7.2.2*	Shipping Manager	When customer requirements are changed, how is this communicated with shipping?	
377	7.2.3.1	Shipping Manager	Does the organization have the ability to communicate necessary information in a customer-specified language and format? *(Auditor to sample and verify Advanced Shipment Notification requirements.)*	

Auditor Name _____   Date of Audit _____

Section 3.6 Production Process

	ISO/TS 16949:2002 Clause * indicates ISO 9001:2000	Function/Level	Question	Result
378	7.5.3*/ Note	Shipping Manager	How is product identified in your areas?	
379	7.5.5*	Shipping Manager	How is product preserved during storage and delivery?	
380	7.5.5*	Shipping Manager	What controls do you have in place for proper handling, packaging, storage, and protection of product?	
381	7.5.5.1	Shipping Manager	What system do you have in place to ensure stock rotation?	
382	8.2.1*/ Note	Shipping Manager	Who are your internal customers and how do you measure their satisfaction?	

Auditor Name _____   Date of Audit _____

Section 3.6 Production Process

	ISO/TS 16949:2002 Clause * indicates ISO 9001:2000	Function/Level	Question	Result
383	8.2.1.1	Shipping Manager	Show me how you track on-time delivery and premium freight.	
384	8.2.2.3	Shipping Manager	What product audits does your area conduct, if any? *(Auditor to sample records and verify defined frequency is met.)*	
385	8.2.3*	Shipping Manager	What do you do when planned results are not achieved? *(Examples of correction and corrective action should be provided.)*	
386	8.2.3*/ 4.1.e*	Shipping Manager	How do you monitor and measure your process? Please provide examples. This should include data pertaining to characteristics and trends of process.	

Auditor Name _____   Date of Audit _____

Section 3.6 Production Process

	ISO/TS 16949:2002 Clause * indicates ISO 9001:2000	Function/Level	Question	Result
387	8.2.4*/ Note	Shipping Manager	How do you make certain all planned arrangements have been satisfactorily completed prior to shipment?	
388	8.3.4	Shipping Manager	How do you identify material shipped on a customer authorization?	
389	8.4.c*	Shipping Manager	What data does your area collect and analyze pertaining to opportunities for preventive actions?	
390	8.5.1*	Shipping Manager	Please provide examples of continual improvement in your area(s) of responsibility.	

Auditor Name _____   Date of Audit _____

Section 3.6 Production Process

	ISO/TS 16949:2002 Clause * indicates ISO 9001:2000	Function/Level	Question	Result
391	8.5.2	Shipping Manager	What internal corrective actions have been issued/assigned to or closed in your area(s) since the last internal audit? *(Auditor should randomly sample at least six corrective actions.)*	
392	8.5.2	Shipping Manager	What customer complaints have been issued/assigned to or closed in your area(s) since the last internal audit? *(Auditor should randomly sample at least six customer complaints.)*	

Auditor Name _____   Date of Audit _____

Section 3.6 Production Process

	ISO/TS 16949:2002 Clause * indicates ISO 9001:2000	Function/Level	Question	Result
393	8.5.2.4	Shipping Manager	How do you handle returned product from the customer?	
394	8.5.3*	Shipping Manager	What preventive actions have been issued or closed in your area(s) since the last internal audit? *(Because this number is usually small, the auditor should review all preventive actions.)*	
395	8.5.2/ 8.5.3*	Shipping Supervisor/ Team Leader	Have you participated in any corrective and/or preventive actions? If yes, please give an example.	
396	5.1.a*	Shipping Supervisor/ Team Leader	How do you communicate the importance of meeting customer, regulatory, and legal requirements to your area?	

Auditor Name _____   Date of Audit _____

Section 3.6 Production Process

	ISO/TS 16949:2002 Clause * indicates ISO 9001:2000	Function/Level	Question	Result
397	5.3.d*	Shipping Supervisor/ Team Leader	What does the quality policy mean to you? How do you ensure the quality policy is communicated and understood?	
398	5.4.1*	Shipping Supervisor/ Team Leader	What are the quality objectives for your area?	
399	5.5.3*	Shipping Supervisor/ Team Leader	What communication processes do you use?	
400	5.5.3*	Shipping Supervisor/ Team Leader	How effective is your quality management system? How do you know?	

Auditor Name _____   Date of Audit _____

Section 3.6 Production Process

	ISO/TS 16949:2002 Clause * indicates ISO 9001:2000	Function/Level	Question	Result
401	6.2.2.d*	Shipping Supervisor/ Team Leader	How do you contribute to the achievement of the quality objectives?	
402	6.2.2.d*	Shipping Supervisor/ Team Leader	What is the importance of your job?	
403	6.2.2.2	Shipping Supervisor/ Team Leader	What training is required for employees in shipping?	
404	6.3*	Shipping Supervisor/ Team Leader	What additional equipment and tools are needed in the shipping area?	

Auditor Name _____   Date of Audit _____

Section 3.6 Production Process

	ISO/TS 16949:2002 Clause * indicates ISO 9001:2000	Function/Level	Question	Result
405	7.2.2*	Shipping Supervisor/ Team Leader	When a customer order is amended, how is this communicated with shipping?	
406	7.2.2*	Shipping Supervisor/ Team Leader	When customer requirements are changed, how is this communicated with shipping?	
407	7.4	Shipping Supervisor/ Team Leader	Which carriers do you use? *(Use Appendix A, Purchasing.)*	

Auditor Name _____   Date of Audit _____

Section 3.6 Production Process

	ISO/TS 16949:2002 Clause * indicates ISO 9001:2000	Function/Level	Question	Result
408	7.5.3*/ Note	Shipping Supervisor/ Team Leader	How is product identified in your areas?	
409	7.5.5*	Shipping Supervisor/ Team Leader	How is product preserved during storage and delivery?	
410	7.5.5*	Shipping Supervisor/ Team Leader	What controls do you have in place for proper handling, packaging, storage, and protection of product?	

Auditor Name _____   Date of Audit _____

Section 3.6 Production Process

	ISO/TS 16949:2002 Clause * indicates ISO 9001:2000	Function/Level	Question	Result
411	8.3.4	Shipping Supervisor/ Team Leader	How do you identify material shipped on a customer authorization?	
412	8.5.2/ 8.5.3*	Shipping Supervisor/ Team Leader	What corrective and/or preventive actions have you participated in?	
413	5.1.a*	Shipping Associate	How does the company communicate the importance of meeting customer, regulatory, and legal requirements to you?	
414	5.3.d*	Shipping Associate	What does the quality policy mean to you?	

Auditor Name _____   Date of Audit _____

Section 3.6 Production Process

	ISO/TS 16949:2002 Clause * indicates ISO 9001:2000	Function/Level	Question	Result
415	5.4.1*	Shipping Associate	What are the quality objectives for your area?	
416	5.5.1*	Shipping Associate	What are your primary responsibilities? *(If not documented, ensure this matches answer from manager.)*	
417	5.5.1*	Shipping Associate	Whom do you work for? *(Ensure this matches answer from manager.)*	
418	5.5.3*	Shipping Associate	How effective is your quality management system? How is this communicated to you?	
419	6.3*	Shipping Associate	What additional equipment and tools do you need to do your job?	

Auditor Name _____   Date of Audit _____

Section 3.6 Production Process

	ISO/TS 16949:2002 Clause * indicates ISO 9001:2000	Function/Level	Question	Result
420	6.2.2.d*	Shipping Associate	How do you contribute to the achievement of the quality objectives?	
421	6.2.2.d*	Shipping Associate	What is the importance of your job?	
422	6.2.2.2	Shipping Associate	What training have you had?	
423	6.2.2.3	Shipping Associate	For the part or material you are shipping, if a nonconformity is shipped or the part is shipped late, what kind of problems would that create for the customer?	

Auditor Name _____   Date of Audit _____

Section 3.6 Production Process

	ISO/TS 16949:2002 Clause * indicates ISO 9001:2000	Function/Level	Question	Result
424	6.2.2.4	Shipping Associate	What process does the company have to inspire you to achieve your quality goals, to make continual improvements, and to encourage new ideas?	
425	7.5.3*/ Note	Shipping Associate	What product are you loading and how is it identified?	
426	7.6	Shipping Associate	For the product that you're loading, what monitoring and measuring devices do you use? *(Record these instruments on Appendix D, Calibration.)*	
427	7.6	Shipping Associate	How do you know this measuring device is acceptable to use?	

Auditor Name _____   Date of Audit _____

Section 3.6 Production Process

	ISO/TS 16949:2002 Clause * indicates ISO 9001:2000	Function/Level	Question	Result
428	7.5.5*	Shipping Associate	Does the product that you're loading (name of product: _____) have any requirements for: Handling? Packaging? Storage? Protection? *(Auditor should verify against customer requirements/order/procedures/work instructions.)*	
429	8.2.4*/ Note	Shipping Associate	For the product that you're loading (name of product: _____), do you monitor and/or measure the product? *(Verify answer is correct and that measurements were taken at appropriate frequency and met requirements.)*	

Auditor Name _____   Date of Audit _____

Section 3.6 Production Process

	ISO/TS 16949:2002 Clause * indicates ISO 9001:2000	Function/Level	Question	Result
430	8.2.4*	Shipping Associate	How do you know that the product being loaded is okay to ship?	
431	8.3	Shipping Associate	What do you do if you find product out of specification? What happens to the product?	
432	8.3.1	Shipping Associate	What do you do with unidentified or suspect product?	
433	8.3.4	Shipping Associate	How do you identify material shipped on a customer authorization?	

Auditor Name _____   Date of Audit _____

Section 3.6 Production Process

	ISO/TS 16949:2002 Clause * indicates ISO 9001:2000	Function/Level	Question	Result
434	8.5.2/8.5.3*	Shipping Associate	Have you participated in any corrective and/or preventive actions? If yes, please give an example.	
435	6.1*	Determine by observation in Shipping	*(Do resources appear to be adequate to maintain the quality system and continually improve its effectiveness? … to enhance customer satisfaction by meeting customer requirements?)*	
436	6.3*	Determine by observation in Shipping	*(Infrastructure: Does the area include the appropriate workspace, equipment, and supporting services such as transport or communication?)*	
437	6.4*	Determine by observation in Shipping	*(Is the area's work environment appropriate to meet product requirements?)*	

Auditor Name _____   Date of Audit _____

Section 3.6 Production Process

	ISO/TS 16949:2002 Clause * indicates ISO 9001:2000	Function/Level	Question	Result
438	4.2.3*	All—Shipping	*(Randomly sample at least eight documents in the area to ensure they meet requirements of document control procedures. Be sure to include the quality manual, procedures, work instructions, forms, and external documents. Use the Appendix B documents form.)*	
439	4.2.4*/ 4.2.4.1	All—Shipping	*(Randomly sample at least four records in the area to ensure they meet requirements of record control procedures. Use the Appendix C records form.)*	

Auditor Name _____   Date of Audit _____

Section 3.6 Production Process

Use the following section to develop your own questions:

	Procedure/Work Instruction	Function/Level	Question	Result

Auditor Name _____ Date of Audit _____

Section 3.7 Preventive and Corrective Process

Interview the person who coordinates the entire corrective and preventive action system. For samples taken, follow up in areas assigned responsibility to verify effectiveness during the audit of that process. Ensure the actions taken were effective.

	ISO/TS 16949:2002 Clause * indicates ISO 9001:2000	Question	Result
1	4.2.3*/ 8.5.2	Where is your procedure for corrective action and how is it controlled?	
2	4.2.3*/ 8.5.3*	Where is your procedure for preventive action and how is it controlled?	
3	4.2.4*/ 8.5.2	*(Sample records of results of action taken for corrective actions utilizing Appendix C, Records.)*	
4	4.2.4*/ 8.5.2*	*(Sample records of results of action taken for preventive actions utilizing Appendix C, Records.)*	
5	5.5.2.1	Who has been designated the customer representative for corrective and preventive actions?	

Auditor Name _____ Date of Audit _____

Section 3.7 Preventive and Corrective Process

	ISO/TS 16949:2002 Clause * indicates ISO 9001:2000	Question	Result
6	8.2.3*	How do you make sure that corrective action is taken as appropriate when planned results are not achieved?	
7	8.2.3.1	When a manufacturing process becomes unstable or is no longer capable, how do you make sure a corrective action is completed?	
8	8.5.2	How do you manage the corrective action process to ensure its effectiveness?	

Auditor Name _____   Date of Audit _____

Section 3.7 Preventive and Corrective Process

	ISO/TS 16949:2002 Clause * indicates ISO 9001:2000	Question	Result
9	4.1.e*/ 8.2.3*	How do you measure the effectiveness of the corrective action process?	
10	8.5.1*	Can you show me evidence of continual improvement in the corrective and preventive action process?	

Auditor Name _____   Date of Audit _____

Section 3.7 Preventive and Corrective Process

	ISO/TS 16949:2002 Clause * indicates ISO 9001:2000	Question	Result

11	8.5.2*	What internal corrective actions have been issued/assigned to or closed in your plant(s) since the last internal audit? *Auditor should randomly sample at least eight corrective actions. Ensure each includes:*
		
		                                       1 2 3 4 5 6 7 8
		  1. Review of the issue
		
		  2. The causes
		
		  3. Evaluation of the need for preventive action
		
		  4. Determination and implementation of the action needed
		
		  5. Review of action taken	

	8.5.2.1	6. Problem-solving method	
	8.5.2.2	7. Error proofing	
	8.5.2.3	8. Impact	

Auditor Name _____   Date of Audit _____

Section 3.7 Preventive and Corrective Process

	ISO/TS 16949:2002 Clause * indicates ISO 9001:2000	Question	Result

12	8.5.2*	What customer complaints have been issued/assigned to or closed in your plant(s) since the last internal audit? *Auditor should randomly sample at least eight customer complaints. Ensure each includes:*
		
		                                       1 2 3 4 5 6 7 8
		  1. Review of the issue
		
		  2. The causes
		
		  3. Evaluation of the need for preventive action
		
		  4. Determination and implementation of the action needed
		
		  5. Review of action taken	

	8.5.2.1	6. Customer's method	
	8.5.2.2	7. Error proofing	
	8.5.2.3	8. Impact	

Auditor Name _____   Date of Audit _____

Section 3.7 Preventive and Corrective Process

	ISO/TS 16949:2002 Clause * indicates ISO 9001:2000	Question	Result
13	8.5.2.4	Where are the records of corrective actions initiated for parts rejected by the customer? Show me.	
14	8.5.3*	How do you manage the preventive action process to ensure its effectiveness?	
15	4.1.e*/ 8.2.3*	How do you measure the effectiveness of the preventive action process?	

Auditor Name _____   Date of Audit _____

Section 3.7 Preventive and Corrective Process

	ISO/TS 16949:2002 Clause * indicates ISO 9001:2000	Question	Result

16	8.5.3*	What preventive actions have been issued or closed in your plant(s) since the last internal audit? *Because this number is usually small, the auditor should review all preventive actions. Ensure each includes:*
		
		                                           1 2 3 4 5 6 7 8
		
		  1. Determine potential nonconformity
		
		  2. Its potential causes
		
		  3. Evaluation of the need for preventive action
		
		  4. Determination and implementation of the action needed
		
		  5. Review of action taken	

Auditor Name _____   Date of Audit _____

Section 3.7 Preventive and Corrective Process

Use the following section to develop your own questions:

Procedure/Work Instruction	Function/Level	Question	Result

Auditor Name _____   Date of Audit _____

Section 3.8 Internal Quality Audit Process

Interview the person who is responsible for the internal audit process. Randomly sample audit records based on the audit plan.

	ISO/TS 16949:2002 Clause * Indicates ISO 9001:2000	Question	Result
1	4.2.3*/ 8.2.2	May I see your procedure for internal quality audits? How is it controlled? *(Auditor to verify that it meets requirements of ISO/TS 16949:2002.)*	
2	4.2.4*/ 8.2.2	*(Sample audit results records utilizing Appendix C, Records.)*	
3	8.2.2	Please show me your annual audit plan. *(Auditor to verify plan meets requirements of ISO/TS 16949:2002 and procedure. Ensure audits have been performed to plan.)*	
4	8.2.2*	Show me how audits were planned based on the status and importance of the processes to be audited as well as the results of previous audits.	

Auditor Name _____   Date of Audit _____

Section 3.8 Internal Quality Audit Process

	ISO/TS 16949:2002 Clause * indicates ISO 9001:2000	Question	Result
5	8.2.2.4	Show me evidence that the audit frequency was increased when internal/external nonconformities or customer complaints occurred.	
6	8.2.2*	Show me how the following were defined: Audit criteria Scope Frequency Methods	
7	8.2.2*	*(For each audit sampled, verify that the auditors did not audit their own work.)*	

Auditor Name _____   Date of Audit _____

Section 3.8 Internal Quality Audit Process

	ISO/TS 16949:2002 Clause * indicates ISO 9001:2000	Question	Result
8	8.2.2*	For nonconformities found, show me that the responsible management ensured actions were taken without undue delay to eliminate problems and their causes.	
9	8.2.2*	For nonconformities found, show me that verification of actions taken and reporting of verification results were included in follow-up activities.	
10	8.2.2	Show me how the audit determined that the system met requirements of the ISO/TS 16949:2002 and company procedure? If it did not, what action will be taken?	
11	8.2.2*	Show me how the audit determined that the system was effectively implemented and maintained? If it did not, what action will be taken?	

Auditor Name _____   Date of Audit _____

Section 3.8 Internal Quality Audit Process

	ISO/TS 16949:2002 Clause * indicates ISO 9001:2000	Question	Result
12	8.2.2.1	Show me that the audit covered the quality management system to verify compliance to ISO/TS 16949:2002 and any other requirements (for example, customer specific requirements).	
13	8.2.2.2	Show me that each manufacturing process was audited for effectiveness.	
14	8.2.2.4	Show me that the audits covered all processes, activities, and shifts and was scheduled per an annual audit plan.	
15	8.2.2.4/ Note	Show me that specific checklists were used for each audit.	

Auditor Name _____   Date of Audit _____

Section 3.8 Internal Quality Audit Process

	ISO/TS 16949:2002 Clause * indicates ISO 9001:2000	Question	Result
16	8.2.2.5	Show me how the internal auditors were qualified to audit the requirements of ISO/TS 16949:2002.	
17	8.5.1	Show me evidence of continual improvement in the internal audit process.	

Auditor Name _____   Date of Audit _____

Section 3.8 Internal Quality Audit Process

Use the following section to develop your own questions:

Procedure/Work Instruction	Function/Level	Question	Result

Auditor Name _____   Date of Audit _____

Section 3.9 Resource Management/Training Process

	ISO/TS 16949:2002 Clause * indicates ISO 9001:2000	Function/Level	Question	Result
1	4.1.a*	HR Manager	What process(es) do you manage?	
2	4.1.b*	HR Manager	How does your process(es) link with the other processes in the company?	
3	5.1.a*	HR Manager	How do you communicate the importance of meeting customer, regulatory, and legal requirements to your area?	
4	5.3.d*	HR Manager	How do you ensure the quality policy is communicated and understood?	

Auditor Name _____   Date of Audit _____

Section 3.9 Resource Management/Training Process

	ISO/TS 16949:2002 Clause * indicates ISO 9001:2000	Function/Level	Question	Result
5	5.4.1*	HR Manager	What are the quality objectives for your area?	
6	5.4.2*	HR Manager	How do you ensure the integrity of the quality management system is maintained when changes are planned and implemented?	
7	5.5.1*	HR Manager	Where are responsibilities and authorities defined for the plant, including HR, and how are they communicated?	

Auditor Name _____   Date of Audit _____

Section 3.9 Resource Management/Training Process

	ISO/TS 16949:2002 Clause * indicates ISO 9001:2000	Function/Level	Question	Result
8	5.5.2.1	HR Manager	Who has the responsibility to ensure customer requirements are addressed in establishing quality objectives and related training?	
9	5.5.3*	HR Manager	What communication processes are established within your area?	
10	5.5.3*	HR Manager	How do you communicate the effectiveness of the quality management system to your area?	

Auditor Name _____   Date of Audit _____

Section 3.9 Resource Management/Training Process

	ISO/TS 16949:2002 Clause * Indicates ISO 9001:2000	Function/Level	Question	Result
11	6.1.a*	HR Manager	How does the company determine and provide the resources needed to maintain and continually improve the effectiveness of the quality system?	
12	6.1.b*	HR Manager	How does the company determine and provide the resources needed to improve customer satisfaction by meeting customer requirements?	

Auditor Name _____   Date of Audit _____

Section 3.9 Resource Management/Training Process

	ISO/TS 16949:2002 Clause * Indicates ISO 9001:2000	Function/Level	Question	Result
13	6.2.1*	HR Manager	How do you determine competence of your employees? Salaried: Hourly:	
14	6.2.2.c*	HR Manager	How do you evaluate training effectiveness? Please show examples as evidence.	

Auditor Name _____   Date of Audit _____

Section 3.9 Resource Management/Training Process

	ISO/TS 16949:2002 Clause * indicates ISO 9001:2000	Function/Level	Question	Result
15	6.2.2.d*	HR Manager	How do you ensure that personnel in HR are aware of the importance of their jobs and how they contribute to achievement of quality objectives?	
16	6.2.2.1	HR Manager	Show me where appropriate tools and methods needed for product design personnel have been identified.	
17	6.2.2.1	HR Manager	How do you make certain that persons responsible for product design are capable to do their job and skilled in appropriate tools and methods identified?	

Auditor Name _____   Date of Audit _____

Section 3.9 Resource Management/Training Process

	ISO/TS 16949:2002 Clause * indicates ISO 9001:2000	Function/Level	Question	Result
18	6.2.2.2	HR Manager	Show me your documented procedure for determining training needs and attaining capability of all employees impacting product quality.	
19	6.2.2.2	HR Manager	How are personnel performing specific assigned tasks qualified?	
20	6.2.2.3	HR Manager	What on-the-job training is given for employees in a new or changed job impacting product quality?	
21	6.2.2.3	HR Manager	Who are the contract or agency personnel you have working? Show me their on-the-job training records.	

Auditor Name _____   Date of Audit _____

Section 3.9 Resource Management/Training Process

	ISO/TS 16949:2002 Clause * indicates ISO 9001:2000	Function/Level	Question	Result
22	6.2.2.3	HR Manager	How do you inform employees that if a nonconformity is shipped, what kind of problems would that create for the customer?	
23	6.2.2.4	HR Manager	What is the company's process to inspire employees to meet quality goals, to continually improve, and to encourage new ideas?	
24	6.2.2.4	HR Manager	How does the process just mentioned include promotion of quality technological consciousness all through the company?	
25	6.2.2.4	HR Manager	What process do you use to measure how much personnel know of the significance of their job and how they add to the attainment of the quality goals? Please show me the results.	

Auditor Name _____   Date of Audit _____

Section 3.9 Resource Management/Training Process

	ISO/TS 16949:2002 Clause * indicates ISO 9001:2000	Function/Level	Question	Result
26	6.4.1	HR Manager	How does the company address product safety and ways to reduce possible risks to personnel, particularly in the design and manufacturing processes?	
27	8.1.2	HR Manager	How do you ensure fundamental statistical concepts are understood and used?	
28	8.2.3*	HR Manager	What do you do when planned results are not achieved? (Examples of correction and corrective action should be provided.)	

Auditor Name _____   Date of Audit _____

Section 3.9 Resource Management/Training Process

	ISO/TS 16949:2002 Clause * indicates ISO 9001:2000	Function/Level	Question	Result
29	8.2.3*/ 4.1.e*	HR Manager	How do you monitor and measure your process? Please provide examples. This should include data pertaining to characteristics and trends of process.	
30	8.4.c*	HR Manager	What data does your area collect and analyze pertaining to opportunities for preventive actions?	
31	8.5.1*	HR Manager	Please provide examples of continual improvement in your area(s) of responsibility.	

Auditor Name _____   Date of Audit _____

Section 3.9 Resource Management/Training Process

	ISO/TS 16949:2002 Clause * indicates ISO 9001:2000	Function/Level	Question	Result
32	8.5.2	HR Manager	What internal corrective actions have been issued/assigned to or closed in your area(s) since the last internal audit? *(Auditor should randomly sample at least six corrective actions.)*	
33	8.5.3*	HR Manager	What preventive actions have been issued or closed in your area(s) since the last internal audit? *(Because this number is usually small, the auditor should review all preventive actions.)*	
34	5.1.a*	Training Coordinator	How does the company communicate the importance of meeting customer, regulatory, and legal requirements to you?	

Auditor Name _____   Date of Audit _____

Section 3.9 Resource Management/Training Process

	ISO/TS 16949:2002 Clause * indicates ISO 9001:2000	Function/Level	Question	Result
35	5.3.d*	Training Coordinator	What does the quality policy mean to you?	
36	5.4.1*	Training Coordinator	What are the quality objectives for your area?	
37	5.5.1*	Training Coordinator	What are your primary responsibilities? *(If not documented, ensure this matches answer from manager.)*	
38	5.5.1*	Training Coordinator	Whom do you work for? *(Ensure this matches answer from manager.)*	

Auditor Name _____   Date of Audit _____

Section 3.9 Resource Management/Training Process

	ISO/TS 16949:2002 Clause * indicates ISO 9001:2000	Function/Level	Question	Result
39	5.5.3*	Training Coordinator	How effective is your quality management system? How is this communicated to you?	
40	6.3*	Training Coordinator	Do you need additional equipment and tools to do your job?	
41	6.2.2.d*	Training Coordinator	How do you contribute to the achievement of the quality objectives?	
42	6.2.2.d*	Training Coordinator	What is the importance of your job?	

Auditor Name _____   Date of Audit _____

Section 3.9 Resource Management/Training Process

	ISO/TS 16949:2002 Clause * indicates ISO 9001:2000	Function/Level	Question	Result
43	8.5.2/ 8.5.3*	Training Coordinator	Have you participated in any corrective and/or preventive actions? If yes, please give an example.	
44	6.2.2*	Training Coordinator	*(Using Appendix G, Interview Sheet, randomly sample employees interviewed. Be certain to sample employees from each process and level in the organization. Ensure each employee has a record indicating competence to do the job he/she was performing while interviewed. In the third column of the Appendix G interview sheet, indicate whether not competency was determined and appropriate employees had been qualified. In the fourth column, indicate whether or not training requirements had been met, including on the job. In the fifth column, check off if the employee is aware of the quality policy and objectives. Don't forget to include the plant manager.*	
	6.2.2.3 4.2.4* 4.2.4.1		*Make certain records maintained indicate appropriate education, training, skills, and experience and meet requirements for record control.)*	

Auditor Name _____   Date of Audit _____

Section 3.9 Resource Management/Training Process

Use the following section to develop your own questions:

Procedure/Work Instruction	Function/Level	Question	Result

Auditor Name _____   Date of Audit _____

Section 3.10 Product Realization Process

To ensure the entire product realization process is working, randomly choose a product that shipped sometime during the previous month.

Product: _____   Customer: _____

Date of production: _____   Date of shipment: _____   Carrier: _____

	Clause	Question	Result
1	7.1/ 7.3	*(Check the records for planning of product realization and design.)*	
2	7.2	*(Obtain the record of the order and any customer-specific requirements. Ensure the order was processed properly. Verify all customer-specific requirements were met throughout the rest of this audit.)*	
3	7.2	*(Verify manufacturing viability had been examined, confirmed, and documented, including risk analysis during the contract review process.)*	

Auditor Name _____   Date of Audit _____

Section 3.10 Product Realization Process

	Clause	Question	Result
4	7.3.6.3	*(Verify that the customer's product approval process had been correctly utilized.)*	
5	7.3.6.3	*(Verify that the customer's product approval process had been applied to suppliers as well.)*	
6	7.4	*(Check records of these raw materials to ensure they were received and inspected as required. Sample any measuring devices that were used by placing them on Appendix D, Calibration.)*	

Auditor Name _____   Date of Audit _____

Section 3.10 Product Realization Process

	Clause	Question	Result
7	7.4	*(Verify that the suppliers of the raw materials and the carrier were approved and had been evaluated and reevaluated.)*	
8	7.4	*(Verify that the suppliers of raw materials were ISO 9001:2000 certified. If not, verify that there is a plan in place for them to get there during the first three-year cycle of the organization's ISO/TS 16949:2002 certificate.)*	
9	7.4	*(Verify that the organization has worked with the suppliers of raw materials to develop them with the goal of conformity to ISO/TS 16949:2002.)*	

Auditor Name _____   Date of Audit _____

Section 3.10 Product Realization Process

	Clause	Question	Result
10	7.4	*(Verify the purchase orders for the raw materials and carrier.)*	
11	7.4	*(Determine whether any processes were outsourced. If so, verify methods of control.)*	
12	7.5	*(Verify production records to ensure product characteristics and process parameters were met per the control plan.)*	

Auditor Name _____   Date of Audit _____

Section 3.10 Product Realization Process

	Clause	Question	Result
13	7.5.1.2	*(Verify that work instructions were available for applicable processes.)*	
14	7.5.1.3	*(Verify that job setups had been verified.)*	
15	7.5.1.4	*(For processes utilized, verify that preventive and predictive maintenance had been performed; verify availability of spare parts.)*	
16	7.5.1.5/ 7.5.4.1	*(For the particular tooling involved with this part, verify its management and history. Verify that tool is permanently marked so that ownership is visible and can be determined.)*	
17	7.5.2	*(Verify that the process had been validated and revalidated.)*	

Auditor Name _____   Date of Audit _____

Section 3.10 Product Realization Process

	Clause	Question	Result
18	7.5.3	*(Verify the order is traceable back to the raw materials. List all raw materials.)*	
19	7.5.4	*(If any customer property was involved, verify requirements of 7.5.4 were met.)*	
20	7.6/ 7.6.1	*(Verify instruments used for monitoring and measuring had been calibrated or verified and met the requirements of 7.6. Verify that gage variation study had been performed for each type of instrument on the control plan.)*	
21	7.6.3.2	*(If an external lab was used to test the product, ensure it was approved by the customer, accredited, or the OEM.)*	
22	8.2.3.1	*(Verify that the manufacturing process capability or performance was being sustained as specified by the customer's product approval process requirements.)*	

Auditor Name _____   Date of Audit _____

Section 3.10 Product Realization Process

	Clause	Question	Result
23	8.2.3.1	*(Verify that significant process events such as tool change or machine repair were recorded.)*	
24	8.2.3.1	*(If the process had become unstable or not capable, verify that the reaction plan had been initiated. Verify that the reaction plan included containment and 100% inspection as appropriate. Verify that corrective action had been taken to make certain the process becomes stable and capable.)*	
25	8.2.1	*(Check the customer's scorecard to determine its satisfaction.)*	

Auditor Name _____   Date of Audit _____

Section 3.10 Product Realization Process

	Clause	Question	Result
26	8.2.4	*(Verify in-process and final inspection records to ensure all tests and inspections were performed and met requirements.)*	
27	8.3	*(Verify that all requirements for nonconforming product were met as needed, including customer waiver.)*	

Auditor Name _____   Date of Audit _____

Section 3.10 Product Realization Process

Use the following section to develop your own questions:

	Procedure/Work Instruction	Function/Level	Question	Result

Auditor Name _____   Date of Audit _____

Conclusion

Your process approach audit checklist will only improve with time as you add more and more questions specific to your organization. I suggest that your organization collect questions from different auditors across different processes to optimize the checklist. The checklist should be a living document, changing and growing with your organization and its auditors.

Auditing is not for everyone, but proper training and the right tools make it much more effective, enjoyable, and easier for those who only audit occasionally. It is my sincerest hope that this manual has left you better prepared as an auditor as well as an auditee. When I took my first auditing class 10 years ago, I never dreamed I would be a professional auditor today. Who knows? It might happen to you.

HAPPY AUDITING!

Appendix A
Purchasing

Raw material	Material code	Lot number	Supplier	Date rec'd	Proper ID?	Inspection completed & material passed?	C of A avail if req'd?	Approved supplier?	Supplier selected properly?	Supplier ISO 2000 certified?	Supplier evaluated?	Supplier reevaluated?	PO adequate and includes QMS requirements?	Adequacy confirmed prior to order?	Product approval process recognized by customer?	Supplier monitored per 7.4.3.2 of ISO/TS 16949:2002
1																
2																
3																
4																
5																
6																
7																
8																
9																
10																
11											ISO 17025 accredited						
12																
Lab/calibration provider																
1	X	X		X	X	X	X									
2	X	X		X	X	X	X									
3	X	X		X	X	X	X									
Carriers																
1	X	X		X	X	X	X			X						
2	X	X		X	X	X	X			X						
3	X	X		X	X	X	X			X						
Other services																
1	X	X		X	X	X	X			X						
2	X	X		X	X	X	X			X						

Appendix B
Documents

Name	Doc #	Rev #/date	Location	Master list rev #/date	Review & approval	OK?
1						
2						
3						
4						
5						
6						
7						
8						
9						
10						
11						
12						
13						
14						
15						
16						
17						
18						
19						
20						
21						
22						
23						

Appendix C
Records

Name	Location	Legible?	Retrievable?	Protected?	Retention time required	Actual time retained	Disposition?	Satisfies customer/ regulatory?
1								
2								
3								
4								
5								
6								
7								
8								
9								
10								
11								
12								
13								
14								
15								
16								
17								
18								
19								
20								
21								

Appendix D
Calibration

Type of instrument	ID number	Location	Last calibrated	Calibration due date	Calibration interval established	In tolerance? If not, impact assessed?	Traceable standard used?	Gage variation study if on control plan	Standard identified on record?	Revision after engr changes	Stated conformity after calibration?	Calibration on lab scope?
1												
2												
3												
4												
5												
6												
7												
8												
9												
10												
11												
12												
13												
14												
15												
16												
17												
18												
19												
20												

Note: If outside calibration service used, enter on Appendix A.

Appendix E
Management Review

Date of meeting: _____

Did review input include: **(Yes or no for each)**

1. Audit results	
2. Customer feedback	
3. Process performance	
4. Product conformity	
5. Status of corrective actions	
6. Status of preventive actions	
7. Follow-up actions from previous management reviews	
8. Changes that could affect the quality system, including the quality policy and quality objectives	
9. Recommendations for improvement	
10. Summary of design measurements (7.3.4.1)	
11. Analysis of actual & potential field failures	

Did review output include decisions and actions related to: **(Yes or no for each)**

1. Improvement of the system's effectiveness and its processes	
2. Improvement of the product in relation to customer requirements	
3. Resource needs	

Other: **(Yes or no for each)**

Were all areas represented at the meeting?	
Was the review conducted within the established time frame?	
Were records maintained as required?	
Was the system found suitable, adequate, and effective?	
Was the policy reviewed for continuing suitability?	
Was the cost of poor quality reported and evaluated?	
Were all requirements of the quality management system included?	
Were performance trends included in the review?	

Appendix F
Customer-Specific
Requirements—
Responsibility Matrix

Customer	Name of customer-specific requirement	Rev date	Page no.	Section no.	Summary of requirement	Responsibility—include names if possible	Audit result

Appendix G
Interview Sheet

The following persons were interviewed and provided evidence:

Print name	Print department and title Auditor should verify that title matches job being done during the audit.	Competence	Training	Policy/ objective?

Appendix H
Management Review
Meeting Checklist

Date of meeting: _____

Meeting attendee	Title
_____	_____
_____	_____
_____	_____
_____	_____
_____	_____
_____	_____

Use a separate sheet if necessary.

Review input

1. Audit results
2. Customer feedback
3. Process performance
4. Product conformity
5. Status of corrective actions
6. Status of preventive actions
7. Follow-up actions from previous management reviews
8. Changes that could affect the quality system, including quality policy/objectives
9. Recommendations for improvement
10. Summary of supplier performance
11. Summary of design measurements (see 7.3.4.1)
12. Cost of poor quality
13. Analysis of actual and potential field failures
14. All requirements of the quality management system
15. Performance trends

Review output

1. Improvement of the system's effectiveness and its processes
2. Improvement of the product in relation to customer requirements
3. Resource needs

Was the system found suitable, adequate, and effective?	
Does the policy continue to be suitable? If no, what changes should be made?	

References

1. Technical Specification ISO/TS 16949 Quality management systems— Particular requirements for the application of ISO 9001:2000 for automotive production and relevant service part organizations, Reference number ISO/TS 16949:2002(E), International Automotive Task Force, ISO 2002, 2nd ed. 2002-03-01.
2. Quality System Requirements QS-9000, Chrysler Corporation, Ford Motor Company, General Motors Corporation, October 1998.
3. ANSI/ISO/ASQ Q9001-2000, *Quality Management Systems—Requirements,* ASQ Quality Press, 2000.
4. IATF Automotive Certification Scheme for ISO/TS 16949:2002, Rules for Achieving IATF Recognition, 2nd ed. for ISO/TS 16949:2002 May 31, 2004, pp. 12–13.
5. IATF Automotive Certification Scheme for ISO/TS 16949:2002, Rules for Achieving IATF Recognition, 2nd ed. for ISO/TS 16949:2002 May 31, 2004, p. 4.
6. ISO/TS 16949:2002(E), p. 2
7. ANSI/ISO/ASQ Q9000-2000, *Quality Management Systems—Fundamentals and Vocabulary,* ASQ Quality Press, 2000, 3.9.12.
8. ISO/TS 16949:2002(E), p. 14
9. Production Part Approval Process (PPAP), Chrysler Corporation, Ford Motor Company, General Motors Corporation, 1999.
10. http://www.iaob.org/faq.html, as of February 25, 2005.
11. ANSI/ISO/ASQ Q9000-2000, *Quality Management Systems—Fundamentals and Vocabulary,* ASQ Quality Press, 2000, 3.9.3.
12. ISO/TC 176/SC 1/N 215, ISO/TC 176/SC 2/N 526R, *ISO 9000 Introduction and Support Package: Guidance on Terminology Used in ISO 9001:2000 and ISO 9004:2000,* p. 6.
13. Advanced Product Quality Planning (APQP) and Control Plan Reference Manual, Chrysler Corporation, Ford Motor Company, General Motors Corporation, 1995.

Index